Emerging Technologies in Second Language Teaching and Learning

Artificial Intelligence, Adaptive Learning, Virtual Reality, and Beyond

Edited by

Shelly Wyatt

University of Central Florida

Baiyun Chen

University of Central Florida

Maria Redmon

University of Central Florida

Series in Education

VERNON PRESS

www.vernonpress.com

In the Americas:
Vernon Press
1000 N West Street, Suite 1200
Wilmington, Delaware, 19801
United States

In the rest of the world:
Vernon Press
C/Sancti Espiritu 17
Malaga, 29006
Spain

Series in Education

Library of Congress Control Number: 2026942121

ISBN: 979-8-2616-0058-9

Cover design by Vernon Press with elements by Magnific.

Table of contents

List of figures

List of tables

Abbreviations

ACTFL	American Council on the Teaching of Foreign Languages
ADDIE	Analysis, design, development, implementation, and evaluation
AI	Artificial intelligence
AIGC	AI-generated content
AIM	Affordable Instructional Materials
AR	Augmented reality
BICS	Basic interpersonal communication skills
CA	Conversational agent
CALL	Computer assisted language learning
CALP	Cognitive academic language proficiency
CDL	Center for Distributed Learning
CEFRE	Common European Framework of Reference of Languages
CERCLL	Center for Educational Resources in Culture, Language, and Literacy
CLT	Cognitive load theory
COVID-19	Coronavirus Disease 2019
CRI	Course Redesign Initiative
CRSE	Culturally responsive and sustaining education/practice
DMLL	Digital mediated language learning
DOF	Degrees of freedom
DFW	Drop, fail, withdraw
EFL	English as a foreign language
ELD	English language development
ELDSF	English language development standards framework
ELED	eLearning Engagement Design
ELL	English language learner
EU	European Union
F2F	Face-to-face

FERPA	Federal Educational Rights and Privacy Act
FFI	Form-focused instruction
FL	First language
GAI	Generative artificial intelligence
GBL	Game-based language learning
GenAI	Generative artificial intelligence
HEA	Higher Education Act
HBO	Home Box Office
HSI	Hispanic-serving institution
ICALL	Intelligent computer-assisted language learning
IBM	International Business Machines
IEP	Individual educational program
IT	Information technology
ITLS	Intelligent tutoring language systems
ITS	Intelligent tutoring systems
K-12	Kindergarten through 12th grade
L1	First language
L2	Second language
LLM	Large language models
MALL	Mobile-assisted language learning
ML	Machine learning
MLL	Modern languages and literatures
MOOC	Massive open online course
MR	Mixed reality
MTSS	Multitiered system of support
NCDPI	North Carolina Department of Public Instruction
NCES	National Center for Education Statistics
NLP	Natural language processing
NPC	Non-player character
OER	Open educational resources
PAL	Personalized adaptive learning

PLC	Professional learning communities
PLD	Professional level descriptors
SAMR	Substitution, augmentation, modification, redefinition
SCT	Sociocultural theory
SDK	Software development kit
SEL	Social-emotional learning
SIOP	Sheltered instruction observation protocol
SLA	Second language acquisition
SLL	Second language learning
SPARC	Scholarly Publishing and Academic Resources Coalition
STEM	Science, technology, engineering, and mathematics
TBLT	Task-based language learning
TePBLL	Technology-enhanced project-based language learning
TESOL	Teaching English to speakers of other languages
TPACK	Technological pedagogical content knowledge
UCF	University of Central Florida
UDL	Universal design for learning
US	United States
USC	United States Code
UTAUT	Unified theory of acceptance and use of technology
VE	Virtual environment
VR	Virtual reality
VRChat	Virtual reality chat
VR-CCL	Virtual reality-constructivism and cognitive load
WEIRD	Western, educated, industrial, rich, and democratic
WIDA	World class instructional design and assessment
XR	Extended reality
ZPD	Zone of proximal development

Introduction

The current state of second language learning in the United States reflects a complex interplay of policy, educational practices, and sociocultural dynamics. Language education, particularly in the context of bilingual and multilingual programs, has gained traction, though it faces significant challenges. For example, despite funding challenges, the U.S. remains a leader in bilingual education research, with a notable increase in scholarly publications over the past decade, suggesting a growing recognition of the importance of bilingualism in educational contexts (Pérez & Manzano-Agugliaro, 2021). The integration of technology into language learning has emerged as a critical factor in enhancing educational outcomes. Digital-based literacy initiatives have shown promise in improving language acquisition among young learners, yet there remains a gap in fully leveraging technology within language-teaching practices. The promotion of equitable access to language education across diverse socioeconomic backgrounds represents an area of opportunity, as disparities in policy implementation can lead to unequal educational experiences (Adetokun et al., 2023; Cruickshank et al., 2020). While the landscape of second language learning continues to evolve, it is marked by significant challenges related to policy support, funding, and the effective integration of technology in educational practices. The ongoing research and advocacy for bilingual education underscore the necessity for a more inclusive and supportive framework that addresses the diverse needs of learners in a globalized world.

The inspiration for this volume arose from the enthusiasm and commitment of our Modern Languages and Literatures faculty to embrace new instructional technologies that promote student success. Foreign language education currently faces a myriad of challenges that hinder its effectiveness and accessibility across various educational contexts. One significant issue is the lack of motivation among students, often exacerbated by insufficient curriculum hours dedicated to language instruction, which can lead to the perception of foreign languages as non-essential subjects within broader educational frameworks (Zablotska et al., 2022). The integration of technology into language learning, while promising, has not been fully realized, as many educators struggle to effectively incorporate digital tools into their teaching methodologies. This gap in technological integration is compounded by the rapid evolution of AI and its implications for language education, where educators must adapt to new tools while ensuring that they enhance rather than detract from traditional pedagogical approaches (Daşkan et al., 2021;

Liang et al., 2023). What's more, the challenge of inclusivity in foreign language education remains a pressing concern. Educators are tasked with accommodating diverse learning needs, including those of students with disabilities, which requires significant adjustments in teaching strategies and resources (Lintangsari & Emaliana, 2020). The psychological aspect of language learning, particularly foreign language anxiety, further complicates the landscape, as many learners experience heightened levels of anxiety that can impede their ability to engage with the language effectively (Li, 2023). Lastly, the need for ongoing professional development for language teachers is critical, as many educators may not feel adequately prepared to meet the evolving demands of foreign language instruction in a globalized context (Chunping et al., 2024). Addressing these challenges is essential for fostering a more effective and inclusive foreign language education system that meets the needs of all learners.

This volume is organized into three sections: Part 1. Foundations of AI and Instructional Innovation in Second Language Acquisition (SLA); Part 2. Research Studies in Emerging Technologies in SLA; and Part 3. Pedagogical Strategies and Reflections for Language Educators.

Part I.
Foundations of artificial intelligence and instructional innovation in second language acquisition

Chapter 1

Implementing AI in second language acquisition: opportunities and challenges

Jess Smith

Bellarmine University

Abstract

This chapter explores how artificial intelligence (AI) technologies intersect with second language acquisition principles and frameworks. It shares specific strategies for how teachers can use AI-powered tools to create more personalized, engaging instruction for multilingual learners while providing scaffolded instruction in a language-rich environment. It begins with an overview of AI and common second language acquisition frameworks and then moves into the strategies and possibilities of using AI through the lens of sociocultural theory. These strategies include using AI for lesson planning, resource creation, and formative assessment, using large language models (MLLs; chatbots) as conversation partners who provide feedback in real-time, and leveraging AI integrations in existing software to take advantage of the processing power. The chapter then moves into a brief exploration of the challenges of using AI in schools and ends with suggestions for the future directions of this work.

Keywords: Artificial intelligence, Generative AI, Second-language acquisition, Multilingual learners, Sociocultural theory

* * *

1. Artificial intelligence, language learning, and the sociocultural turn

In a world increasingly shaped by globalization and technological innovation, the ability to navigate multiple languages has become essential. At the same time, artificial intelligence (AI) is transforming nearly every aspect of education, including the teaching and learning of second languages. For multilingual learners (MLLs), who represent over 10% of U.S. public school students (National Center for Education Statistics [NCES], 2024), AI offers unprecedented opportunities for differentiated instruction, real-time feedback, and scaffolded learning in ways that were previously unimaginable. However, the use of AI in education, particularly with MLLs, also raises essential pedagogical and ethical questions: what do learners gain when AI becomes a regular part of language classrooms? What might learners lose?

This chapter approaches these questions through the lens of sociocultural theory (SCT), which emphasizes the social and mediated nature of learning. From this perspective, language acquisition is not simply a cognitive task, but a fundamentally interactive and collaborative process. Learners construct meaning through dialogic engagement, scaffolded support, and participation in culturally and linguistically rich classroom communities (Vygotsky, 1978). This chapter argues that thoughtfully implemented AI tools can serve as mediational means that extend MLLs' zones of proximal development (ZPD; Vygotsky, 1978), providing access to real-time feedback, structured interaction, and differentiated pathways for learning that would otherwise be unavailable in many classrooms.

Rather than offering a general endorsement of AI, this chapter argues that the value of AI tools lies in their alignment with existing pedagogical frameworks. Specifically, the chapter maps AI tools onto three foundational frameworks in second language instruction: teaching English to speakers of other languages (TESOL), sheltered instruction observation protocol (SIOP), and basic interpersonal communication skills (BICS) and cognitive academic language proficiency (CALP). In doing so, it illustrates the ways AI can support, but not replace, human teachers in their efforts to create inclusive, dialogic, and rigorous language learning environments. The chapter shares classroom strategies to give teachers practical guidance. It also acknowledges the challenges, limitations, and ethical issues involved.

2. Theoretical and pedagogical foundations

Grounding AI in SCT clarifies that technologies matter insofar as they mediate interaction, scaffold participation within learners' ZPD, and sustain culturally situated meaning-making. Intentional, responsible use of AI has a significant possibility to both improve and enhance teaching and learning (Chen & Tsi, 2023; Harry, 2023; Younas et al., 2023) but also carries with it challenges for misuse and bias (Dignum, 2021; Harry, 2023; Lampou, 2023). In particular, AI presents transformative potential in second language acquisition (SLA), aligning with established principles such as those outlined in TESOL, SIOP, and BICS/CALP. This brief review of the scholarship unfolds in three steps: first, an overview and history of AI, second, a short description of SCT, and finally, an exploration of MLLs and second language acquisition (SLA).

2.1. Artificial intelligence

I begin with a brief overview of what AI is and how current generative AI (GenAI) can serve as a tool to mediate effective SLA. Large language models (LLMs) represent a significant milestone in AI's evolution. These models leverage deep learning techniques to process and generate human-like text. The development of LLMs began with simple models, but the models' complexity and capabilities grew alongside computational power and data availability (Brown et al., 2020). The trajectory of AI continues to be dynamic, increasingly used in industry, education, and daily lives. One opportunity is its use in SLA.

The history of AI and LLMs is marked by rapid advancement and transformative innovation and has been tracked by multiple researchers (Baber et al., 2023; Hadi et al., 2023; Raiaan et al., 2024). The concept of AI dates to ancient myths and legends (Duca, 2023), but it was not studied as a viable discipline until the mid-20th century. In 1956, the Dartmouth Conference marked the formal birth of AI as a discipline (McCarthy et al., 1956). The initial years saw the development of symbolic AI, which relied on manually crafted rules and logical reasoning (Newell & Simon, 1976). The 1980s and 1990s introduced the concept of machine learning, a paradigm shift that emphasized learning from data rather than relying solely on pre-programmed rules (Mitchell, 1997). Neural networks, inspired by the human brain's structure, became more prominent during this time (Rumelhart et al., 1986). However, researchers point to the advent of deep learning in the 2010s as the real beginning of the AI revolution (LeCun et al., 2015). Deep learning, driven by large datasets and powerful computational resources, enabled the creation of more sophisticated models capable of recognizing patterns and making predictions with remarkable accuracy (LeCun et al., 2015).

To maximize AI's potential in education while mitigating its risks, educators must be equipped with the knowledge to critically and ethically integrate these tools into their classrooms. Recent research emphasizes the need for language educators to understand and effectively utilize AI and LLMs in teaching. These technologies offer opportunities to enhance classroom instruction, streamline workloads, and provide differentiated learning experiences (Bonner et al., 2023; Tan et al., 2023). Representative LLMs (e.g., ChatGPT, Gemini, Copilot, Claude) illustrate the class of tools discussed here; to avoid product enumeration, subsequent sections refer to these tools only as LLMs.

2.2. Sociocultural theory

SCT emphasizes learning as fundamentally shaped by social interaction and cultural context. Rather than considering learning as simply an individual cognitive process, SCT identifies the core constructs of mediation, interaction, and ZPD as integral to authentic learning (Vygotsky, 1978). In examining these mediating factors, SCT allows exploration of learning as a socially situated, dialogic process—one that lends itself to SLA in natural ways.

At the heart of SCT is the concept of ZPD, which describes the distance between what a learner can do independently and what they can accomplish when guided by a more knowledgeable mentor, whether a teacher, a peer, or even a resource such as AI. Teachers and researchers alike emphasize that the most effective learning occurs within this zone, providing just enough support and scaffolding to stretch a learner's current capabilities and move them toward independence over time (Mirzaei & Eslami, 2015; Sidek, 2011).

Another key tenet of SCT—mediation—involves the use of symbolic tools, like language, and material tools, like manipulatives or technology, to shape and extend cognitive processes. Thoughtful use of GenAI can serve in this mediating capacity, providing scaffolding and supporting metacognitive reflection or participation in language-rich tasks that might otherwise be outside the reach of MLLs.

Interaction plays a vital role in SCT. Real, authentic learning emerges from meaningful social engagement, and chatbots do not provide this, regardless of how skilled their output may be. Dialogue, feedback, questioning, and collaboration are among the most impactful strategies in classroom communities. These interactions help learners internalize linguistic structures and the social and cultural norms of communication. Teachers' and students' interactions are important and cannot be replicated or replaced by GenAI tools. However, when intentionally applied, these tools can augment effective practices through mediation: for example, by facilitating additional practice through simulation or enhancing collaboration across language barriers.

In the context of second language instruction, SCT reminds educators to prioritize learning environments that are socially engaging, culturally responsive, and strategically scaffolded. When aligned with these principles, AI has the potential to become a powerful mediational tool, enhancing, rather than replacing, the human relationships and interactions at the core of language development.

2.3. Multilanguage learners and second language acquisition

With the mechanism (AI) and the lens (SCT) in view, the chapter now turns to the learners and contexts in which these tools operate. According to National Center for Education Statistics (NCES) data, the percentage of U.S. public school students identified as MLLs grew by 1.2 percentage points nationally between 2011 and 2021. In the fall of 2011, 9.4%, or 4.6 million students, were MLLs, a number that grew to 10.6%, or 5.3 million students, by the fall of 2021.

These numbers range from 0.8% in West Virginia to 20.2% in Texas, but in 13 states, MLLs represent more than 10% of all public-school students. This population increased in 38 states while decreasing in the other 12. For instance, Rhode Island's population of MLLs increased by 6.9 percentage points over the past 10 years, while Nevada's population of MLLs decreased by 4.3 percentage points (NCES, 2024), which underscores the need to prepare teachers to provide high-quality second language instruction.

Effective SLA promotes communication and cultural understanding in increasingly diverse classrooms. For decades, SLA has been proven beneficial for learners of all ages and proficiency levels in various learning environments and on different types of assessments (Long, 1983). It allows educators to equip students with the skills necessary to engage in a globalized world, fostering empathy and cross-cultural connections (Pemberton, 2024). Moreover, usage-based instruction encourages a more context-driven approach, which is more effective than traditional methods that often overemphasize spoken production without sufficient theoretical grounding (Pemberton, 2024). Effective instruction centers on both the target language's structure and its meaningful use, with a particular emphasis on interlanguage grammar acquisition (Lightbown, 2014). Recent research underscores the importance of self-regulated, differentiated learning (Zheng & Stewart, 2024) and application-based practice (Pemberton, 2024), which AI can deepen and make more feasible.

Additionally, supporting students' first-language knowledge can contribute to their long-term academic and second-language success (Krashen, 1982; Lightbown, 2014). Writing instruction is particularly crucial for MLLs, as it is a vital skill for academic and professional success (Hyland, 2022). Strategy instruction can also be key, enabling teachers to guide learners to become more effective and efficient in using the target language (Plonsky & Sudina, 2021). However, the effectiveness of instructional approaches is context-specific and hinges on teachers' perspectives and identity (Hyland, 2022) as well as their comfort and skill (Choi et al., 2022).

Successful SLA is crucial for developing essential skills. SLA has been linked to improved academic performance, enhanced cognitive abilities, and advanced first-language literacy skills (Parks et al., 2023); it promotes writing proficiency (Hyland, 2022) and social-emotional learning (SEL) skills, such as fostering empathy, diversity awareness, and social justice (Parks et al., 2023). Furthermore, SLA enhances social cohesion by helping to integrate immigrant populations and support multiculturalism, particularly relevant in countries like the United States, which have significant immigrant populations (Pemberton, 2024). Given these benefits, it is essential to ensure high-quality SLA, which includes attracting and retaining skilled language teachers (Parks et al., 2023). Proficiency in more than one language enhances job opportunities and competitiveness in various disciplines, including business, healthcare, and diplomacy (Pemberton, 2024). Thus, SLA is critical in contemporary educational contexts, contributing not only to individual growth but also to better societal and economic realities. This chapter draws on three foundational frameworks in SLA: TESOL's standards, the SIOP model, and Cummins's (1979) BICS/CALP distinction as lenses through which to consider the affordances and limitations of AI tools in language learning.

3. Artificial intelligence and instructional frameworks

Viewed through the lens of SCT, TESOL, SIOP, and BICS/CALP, these frameworks become design blueprints for using AI to mediate interaction, scaffold language, and apprentice learners into academic discourse communities. This section outlines opportunities for AI through these principles and frameworks in broad terms. The subsequent section, "Opportunities in Implementing AI in SLA," provides more detailed information on implementing these strategies.

3.1. TESOL

TESOL encompasses diverse methodologies, such as communicative language teaching and task-based learning, designed to meet the unique needs of multilingual learners (MLLs). Foundational theories, such as Krashen's (1982) input hypothesis and Brown's (2000) principles of language learning, emphasize comprehensible input, interactional competence, and low-anxiety environments as critical to SLA. Within TESOL, AI can facilitate comprehensible input and interactional practice while reducing affective barriers, provided that tasks remain dialogic and teacher guided. Although AI lacks the multimodal and sociocultural richness of face-to-face instruction, its scalability and responsiveness make it a valuable tool for supporting TESOL's emphasis on individualized, culturally responsive instruction. If teachers use AI tools intentionally, it can help ensure that each student receives what they need to grow. This kind of design support is crucial; SIOP specifies where and how that mediation should occur.

3.2. SIOP

The SIOP model integrates content and language instruction through eight components, including comprehensible input theory, interaction theory, and review & assessment theory (Echevarria et al., 2008). Its components operationalize SCT by structuring scaffolded, content-rich interaction in authentic, relationship-centered ways. While effective, its complexity can overwhelm teachers and students (Hall, 2019). Another criticism of SIOP is its perceived emphasis on surface-level language skills, such as vocabulary and sentence structure, at the expense of more profound cognitive and academic development (Harper & de Jong, 2009). AI can help teachers in planning, implementing, and assessing SIOP. Beyond SIOP, MLLs need mediated practice across both social and academic registers.

3.3. BICS and CALP

Cummins (1979) distinguishes between BICS, everyday conversational skills, and CALP, the more complex language proficiency needed for academic success. Teachers can use AI tools to help students accelerate BICS acquisition by simulating real-world scenarios, such as ordering food or engaging in casual conversations, alongside more traditional practices like turn-and-talks. LLMs provide MLLs with low-stakes opportunities to practice conversational skills and receive instant corrections, which is particularly beneficial in contexts where class sizes limit individualized feedback (Hwang et al., 2022). These low-stakes interactions provide meaningful opportunities for learners to practice and refine their social language skills while receiving real-time corrections and feedback (Fu & Li, 2022). Gamification further enhances BICS learning, as students engaging with interactive, AI-driven tasks report greater motivation and persistence (Garland, 2015; Nathan & Hashim, 2023).

Developing CALP, however, requires more targeted strategies, such as explicit vocabulary instruction and contextualized practice. While traditional methods, such as flashcards and foldables (Zike, 2008), remain effective (Nation, 2001), AI expands possibilities for deeper engagement, supporting CALP acquisition through strategies and tools like spaced repetition and gamification. These alignments suggest concrete opportunities and boundaries for implementation; the question becomes how to do so responsibly.

4. Opportunities in implementing artificial intelligence in SLA

Implementation succeeds when AI augments human mediation (e.g., extending feedback, access, and participation) without automating away the dialogic core of language learning. Implementing AI in SLA presents both opportunities and challenges, as it addresses needs in personalization, accessibility, and equity. Grounded in SLA principles, AI enhances the learning experience by adapting to individual proficiency levels, creating culturally relevant resources, and supporting multilingual and cross-cultural communication. These opportunities allow educators to rethink traditional approaches, leveraging data-driven insights to optimize instructional practices and empower MLLs. However, while AI is a powerful tool, its practical use requires careful planning, teacher expertise, and critical reflection to ensure it complements rather than replaces human instruction.

The following strategies translate SCT into practice by choreographing AI, teacher moves, and peer interaction to keep learning social, scaffolded, and culturally responsive. AI can be a powerful tool for SLA when teachers use it strategically and purposefully. This section describes practical, specific ways teachers can apply AI for planning, implementing, assessing, and community contexts. When appropriate, the chapter lists more than one tool to illustrate applicability across platforms and to support sustained relevance as technology continues to evolve. The named tools are not exhaustive.

4.1. Planning

In planning, teachers can conceptualize GenAI as a planning partner with the ability to generate drafts of scaffolded tasks, language objectives, and materials for teachers to adapt, correct, and implement. AI can serve as a mediating tool to help teachers streamline preparation by providing initial ideas for creating tailored resources, designing lessons, and aligning activities with SLA theories and TESOL standards, all within the context of teacher expertise, student interest, and proficiency.

When designing lessons and curricula, teachers can utilize GenAI to create scaffolding for existing lessons that helps them better meet the diverse needs of their students in the classroom, blending state standards, accommodations for their specific learners, and the cultural context appropriate for their students in the moment. Teachers can use AI tools to generate vocabulary lists, sample dialogues, grammar exercises, or individualized reading passages tailored to specific themes, interests, and proficiency levels. AI can also help teachers differentiate more broadly by suggesting ways to modify lessons for varying proficiency levels and addressing individual needs, such as multilingual learners who are also gifted, struggling students, or combinations of the above, thereby better allowing teachers to meet the needs of twice-exceptional students.

Beyond individual preparation, AI also supports collaborative planning. Teams of teachers can use GenAI to facilitate brainstorming for interdisciplinary projects or develop team-teaching ideas. By pooling insights across subject areas, teachers can co-create units that integrate literacy practices with science, history, or math, which SLA teachers use to help MLLs see meaningful connections between disciplines (DelliCarpini, 2009). AI's capacity to generate multiple drafts or variations of project outlines allows teams to experiment with different approaches before committing to final plans. This collaborative use of AI also aligns with professional learning communities (PLCs), where teachers share strategies and adapt resources to fit their local contexts (Schneider et al., 2012).

AI can help teachers create comprehensible input in real time or support interaction through collaborative tools. For instance, teachers can utilize AI to support planning and make in-the-moment lesson adjustments. GenAI, such as an LLM or education-specific software like Magic School AI, can serve as a thinking partner, offering initial ideas for solutions by simplifying lesson planning and delivery; adapting existing lessons to incorporate SIOP strategies; helping teachers systematically scaffold instruction and reduce cognitive overload for MLLs; and adding SIOP-aligned supports within students' ZPD. While potentially helpful, these tools require teacher oversight and adaptation to bring value to SLA classrooms (Setyaningsih et al., 2024). For educators, the intentional and critical use of AI serves as a valuable resource for designing collaborative activities that address both linguistic and cultural objectives, ensuring that students are included and represented in the learning process.

4.2. Implementation

AI can scaffold, personalize, and widen access within learners' ZPD without replacing teacher expertise; used this way, it becomes a scaffold within a broader ecosystem of support. AI's ability to adapt to individual learners makes it a powerful tool for personalization, aligning with best practices in SLA. LLMs can tailor language input to learners' proficiency levels, reinforcing key concepts while gradually introducing new challenges (Hwang et al., 2022), thereby helping learners retain and apply conversational or advanced academic terminology.

For example, teachers can use AI to scaffold instruction by generating sentence stems or modeling language structures that support MLLs' understanding of academic concepts. These features address SIOP components such as Comprehensible Input and Practice/Application, providing learners with tools to bridge surface-level proficiency and deeper academic understanding. Furthermore, AI-powered simulations, such as those created with an LLM, enable MLLs to engage in context-specific academic discourse, enhancing their ability to navigate challenging linguistic and cognitive tasks in alignment with CALP development.

Teachers can design prompts for LLM chatbots for intentionally selected students to interact with in structured ways, such as engaging in role-plays, responding to questions using sentence stems, or following multimodal prompts individually, while the teacher works with another group of students to orchestrate peer interaction and feedback. Because teachers can create context-specific scenarios for these simulations (e.g., job interviews, team meetings, casual conversations), multilingual learners have the opportunity to practice shifting register, audience awareness, and purpose across settings. To bridge BICS to CALP, one task could involve MLLs first rehearsing with an LLM-mediated scenario that offers graduated prompts, progressing from hints to stems and later to modeled turns. They would then perform the task in small groups with teacher-orchestrated feedback (e.g., revoicing, probing, and elaboration). Finally, MLLs reflect by describing the academic language moves used and planning what to try next time. Sequences like this position AI as a mediating resource while shifting regulation from other- to self-regulation, while keeping teacher-student and student-student interactions at the core of learning.

AI also has the potential to help teachers bridge gaps in accessibility, offering learners opportunities to practice language skills anytime and anywhere with Internet access (Bulathwela et al., 2024). For students with access, this flexibility ensures at-home resources tailored to their needs, even after school hours when in-person or specialized support is unavailable. By supporting multiple languages simultaneously, AI provides MLLs with materials in their *L1* to help them build foundational knowledge and navigate complex academic concepts, though without a mentor to guide them, students may not realize when the AI shares false or misleading information. This support is especially important for fostering CALP, as engaging with difficult content in a familiar language reduces cognitive load, allowing learners to focus on mastering new academic vocabulary and concepts (Cummins, 1979).

Teachers can also use AI to enhance teacher-student interactions through dynamic translation software (e.g., helping each person communicate primarily in their L1*)* or by providing sentence stems in advance of class to support students in becoming more comfortable with *L2*. Translation tools like Microsoft Translator and Google Translate can further enhance accessibility by enabling cross-linguistic collaboration and can be somewhat more trustworthy than other GenAI. These tools allow teachers to communicate more effectively with multilingual learners and their families, fostering a sense of inclusion and belonging in learning environments. For instance, an MLL struggling to grasp academic content in English can use AI translation to access the material in their *L1*, bridge understanding, and then transition back to the target language—if they have access to a device with the app. This approach aligns with the SIOP model's emphasis on scaffolding and building background knowledge, as it ensures that students have the necessary tools to comprehend and engage with content (Echevarria et al., 2008). By lowering barriers to access, teachers can use AI as a tool to work toward equity for learners who might otherwise be excluded. A deeper exploration of access appears in the challenges section.

Moreover, teachers can use AI tools alongside various dialects and linguistic nuances to increase cultural and linguistic inclusivity, a critical need in linguistically diverse classrooms. GenAI tools can also be customized to align with culturally relevant pedagogies, ensuring that learners see their identities and experiences reflected in the learning process (McCallum, 2024). This inclusivity supports both SEL and academic achievement, as learners feel more empowered and confident in their abilities (Krashen, 1982).

AI fosters collaborative learning across linguistic and cultural barriers, promoting equity in SLA by creating opportunities for interaction and inclusion. Tools like Google Docs with AI integration enable MLLs to collaborate in real time, providing immediate feedback on grammar, vocabulary, and academic language use. These features support language development while empowering students to contribute meaningfully to group projects, which can foster a sense of belonging and agency. Collaborative learning environments enabled by AI align with SEL principles, as they encourage teamwork, empathy, and communication across cultural differences (Nair et al., 2012). AI can help teachers adapt learning strategies across languages, reframing cultural narratives and fostering greater cultural awareness (Zheng & Stewart, 2024). Teachers can utilize AI to support students who speak multiple languages and dialects, even if the teacher lacks fluency in those languages. Additionally, by integrating cultural nuances into tasks, AI can enhance cultural competence, enabling learners to navigate diverse social and professional contexts effectively (McCallum, 2024), although teachers should be intentional to avoid bias, tokenism, and stereotypes that can crop up in AI (Makhortykh et al., 2021; Selwyn, 2024; Weber, 2020). In this way, AI supports both the cultural and collaborative dimensions of SLA by fostering a sense of belonging, teamwork, and identity affirmation.

The incorporation of AI tools provides teachers and students with a wide range of opportunities for practice and extension. Teachers can employ AI tools that incorporate game-based retrieval practice for vocabulary (e.g., Duolingo's adaptive learning, Quizlet, Kahoot!, or similar gamified activities) to keep students engaged with interactive, low-stakes learning. Apps like Grammarly or Google Lens can provide real-time language feedback during writing or speaking activities, helping students self-correct without teacher or peer intervention. Teachers can also use translation tools cautiously to help MLLs understand complex instructions while modeling language growth. Students can use tools like ELSA Speak or YouGlish to hear and practice accurate pronunciation with immediate AI-driven feedback.

As teachers plan and deliver instruction, they can integrate AI with learning management systems to enhance the learning experience. Tools like Blackboard, Canvas, or Moodle can integrate with AI to help teachers provide more personalized instruction by responding to individual student needs (.e.g., assigning specific practice, providing additional reading, or even offering some low-stakes feedback) though using AI for grading and higher-stakes feedback introduces ethical concerns (see discussion in Greene Nolan et al., 2024 and the Challenges section of this chapter).

For elementary and middle school students, tools like Newsela and CommonLit can curate culturally relevant reading materials aligned with their interests, age, and reading proficiency. Teachers can refine these resources by prompting an LLM to adjust reading levels or generate comprehension questions. In high school settings, MindMeister and Miro AI can be utilized for creating graphic organizers, enabling students to break down complex ideas and visualize their connections. For example, a teacher might ask students to map themes in a literary text using AI-generated suggestions for connections and key points, drawing students' attention to errors or shortcomings of the tool.

In writing tasks, teachers can use an LLM to generate sentence stems tailored to specific academic language needs, such as constructing arguments or explaining scientific processes. Middle and high school teachers can also leverage Khan Academy's AI tutor for step-by-step scaffolding in subjects such as math or science, helping students build language skills while learning academic content. For collaborative learning, tools like Google Docs, with AI-enabled suggestions, allow students to work in multilingual teams, offering real-time grammar corrections, vocabulary suggestions, and translation features. These strategies enhance classroom engagement, make input more comprehensible, and support differentiated learning for MLLs.

In addition, AI can mediate more surface-level approaches to vocabulary instruction by providing immediate feedback on vocabulary and grammar while freeing teachers to engage students in higher-order thinking tasks. For example, MLLs can chat with LLMs and receive targeted corrections, enabling teachers to devote more class time to fostering academic discourse and critical thinking (Fu & Li, 2022; Yi, 2021). AI-informed vocabulary tools, such as Quizlet, Duolingo, and Memrise, integrate spaced repetition and contextual exercises, helping students internalize academic language more effectively by working within their ZPD, a practice that decades of research support as effective for retaining and applying conversational or advanced academic terminology (Nation, 2001). AI aligns closely with TESOL principles by offering adaptive, individualized input tailored to learners' proficiency levels (Hwang et al., 2022). Branched storytelling software, such as Twine, can help students engage with interactive, nonlinear storytelling in a structured manner, applying new vocabulary in context at specific proficiency levels. Furthermore, this type of low-stakes practice may reduce learner anxiety, supporting Krashen's (1982) affective filter hypothesis. Taken together, AI-informed tools can enrich SLA classrooms by blending gamified, adaptive, and creative practices with established pedagogical principles when implemented intentionally by skilled SLA educators.

The combination of collaborative tools, immersive simulations, and cross-linguistic tools makes AI a valuable resource for fostering collaboration and equity in SLA. By creating opportunities for MLLs to engage more independently and authentically with peers, they can develop language within its social context. By leveraging these capabilities, teachers can better meet the needs of their students.

4.3. Assessment

Proponents of AI assessment argue that its ability to aggregate and analyze data across large groups allows for the identification of broader trends that can inform school-wide or district-level decisions. Others raise concerns about data privacy and the accuracy of analysis (Barnes & Hutson, 2024; Shukla et al., 2025). Due to the numerous criticisms of AI used in summative assessment, this section prioritizes practical strategies for formative assessment and scaffolded reflection. Concerns related to automatic essay scoring and AI summative feedback are explored at more length in the challenges section that follows.

One of AI's most pedagogically significant affordances is its potential to mediate the instructional process through formative assessment data, thereby supporting teacher decision-making. Such insights can help educators better understand learners' needs as they emerge within the ZPD. For example, when using AI tools to track patterns in students' language production, such as recurring errors in syntax or vocabulary use, teachers can determine more appropriate methods to scaffold instruction. Additionally, data collected by AI-driven formative assessment tools, such as those integrated into Khan Academy or Century Tech, can reveal disparities in performance among different student demographics for students using these tools. This information can guide interventions to address equity gaps, helping teachers ensure that MLLs receive the necessary resources and support to thrive academically (Bulathwela et al., 2024). These tools do not replace teacher judgment but serve as a secondary layer of feedback that supports teacher expertise and decision-making. Applying these strategies critically enables teachers to use AI as a scaffold rather than a shortcut, applying the tools as they would any other mediating tool. These insights align with the principles of differentiation and scaffolding found in the SIOP model, where instruction is adjusted to ensure both comprehensibility and rigor (Echevarria et al., 2008).

Teachers can also use AI to scaffold student reflection on their own learning. Metacognitive tools like Reflectly and Vibe offer learners opportunities to engage in self-reflection, tracking their confidence, progress, and retention over time. These platforms provide students with prompts to reflect on their learning experiences, while AI analyzes their responses to identify emotional or cognitive trends that may impact learning outcomes. For example, students can log challenges and successes after each assignment, while the AI suggests tailored strategies for improvement. By analyzing patterns in student performance, teachers can identify common challenges and adjust instruction to meet the needs of the class or individual learners. These AI-mediated tasks not only save time but also empower students to take ownership of their learning, fostering deeper engagement and sustained progress. Decades of research emphasize the importance of metacognition in SLA, as it fosters learner autonomy and helps students internalize their progress and challenges (Kalaja, 1995; Kalaja et al., 2017; Simard, 2004). In this way, AI can enhance both teacher mediation and learner reflection during the assessment process.

By offering real-time feedback and tracking student performance longitudinally, AI tools empower both learners and educators to engage in continuous improvement. Platforms like Grammarly for Education and AI-enabled writing tools enable ongoing formative assessments, allowing students to refine their language use over time while providing teachers with insights into areas that require additional reinforcement. These data-driven insights provide a foundation for evidence-based teaching practices that enhance learning outcomes, especially for diverse and multilingual populations (Hannah et al., 2023; Hopfenbeck et al., 2023).

However, equity and access remain core concerns; mediation must be coupled with inclusion so that AI remains a tool for empowerment rather than exclusion. AI tools can streamline formative assessment, providing actionable feedback and supporting student reflection at a variety of grade and proficiency levels. For younger students, Quizlet and Memrise offer AI-driven flashcards with spaced repetition and games to reinforce vocabulary acquisition. For middle and high school students, an LLM can act as a virtual reviewer during peer review or the source of exemplar texts designed to model specific skills or errors. Teachers can instruct students to paste drafts into these tools and prompt them for feedback on grammar, sentence structure, and academic language use, allowing students to refine their writing before formal teacher evaluations—teaching both writing and more ethical use of GenAI. To assess oral proficiency, tools like Speechace can evaluate pronunciation and fluency, providing students with detailed feedback that they can use to improve. Together, these formative tools highlight AI's potential to provide low-stakes, continuous assessment opportunities that benefit both students and teachers.

Plug-ins such as Grammarly or Read&Write, among others, provide real-time feedback on writing skills and reading comprehension. These tools can highlight places MLLs should rephrase for clarity or simplify sentences. They can also help students recognize correct grammatical structures in this and future writing, increasing MLL comprehension and written expression of academic language. However, critics question the scaffolding of these tools and how students will move beyond their use to write more independently (Fathimatuzahro & Rizkiyah, 2025). Adaptive and automated assessments offer efficiency and speed, which can be attractive to teachers; yet, they must be applied critically and cautiously to preserve fairness and completeness, emphasizing mediation and teacher expertise over automation.

4.4. Community contexts

Community engagement is essential to SLA and SCT, as both frameworks emphasize that language learning emerges from social interaction within broad cultural contexts. Teachers can utilize AI tools to foster connections among students, families, and communities, thereby supporting SLA beyond the classroom. Translation apps can facilitate communication with non-English-speaking families, enabling them to engage more effectively with their children's education. Communication tools facilitate families' participation in the learning community; however, for grading, placement, or health/safety notices, these notes require human review; full automation carries too great a risk of miscommunication. In that way, teachers can treat translations as aids and starting points rather than official records. By positioning AI translation as a supportive tool rather than a replacement, educators strengthen family-school partnerships while minimizing risks.

Students can use AI-powered language-learning apps like Duolingo, Rosetta Stone, or Babbel for independent practice at home with their families—if they have access to the Internet and these apps, which exist at various price points. Students can also use LLMs to clarify homework instructions or practice writing outside of class, with teacher guidance on ethical use and conversations about hallucinations and other common AI concerns (see discussion in North Carolina Department of Public Instruction [NCDPI], 2024). These applications extend SLA opportunities into the home, allowing families to become partners in learning when provided with equitable access and clear guidance.

Educators can harness the power of AI to connect students with local cultural or linguistic communities, matching them with pen pals or community mentors through AI platforms, a new twist on an established practice. Teachers can also use AI-mediated tools like TalkAbroad or similar platforms to enable students with heavy supervision to practice their target language with native speakers in other countries. Through these practices, AI can help expand community engagement beyond the school walls, linking learners with authentic cultural and linguistic contexts that enrich SLA.

4.5. Considerations for application

Across contexts, the guiding questions remain: who benefits when AI becomes a regular part of language classrooms? What might be lost? As AI becomes increasingly integrated into educational spaces, pedagogical intention and theoretical grounding must inform second language instruction. When approached through the lens of SCT, AI tools can meaningfully extend the work of human teachers by functioning as mediational tools that support the development of multilingual learners within their ZPD. By scaffolding academic language through supports such as sentence stems and visual organizers, and by providing structured opportunities for interaction and reflection, AI can help facilitate more dialogic, responsive learning environments, particularly when used in conjunction with teacher-led instruction and other traditional tools.

Significantly, aligning AI tools with established frameworks (e.g., TESOL, SIOP, and BICS/CALP) enhances the benefits to learners when used to augment, but not replace, the social dimensions of learning. SLA, particularly when viewed through an SCT lens, is a deeply relational process. Thus, while AI can offer timely, low-stakes feedback in informal contexts, access to authentic input through teacher-guided prompting, and tools for teacher-mediated differentiation, it is the human educator who creates the classroom culture, facilitates interaction, and guides learners in making sense of the material. These classroom moves point to broader guidance for teachers, leaders, and researchers.

5. Challenges with implementing artificial intelligence: when mediation becomes automation

Equally important are the limits and risks when mediation slides into automation and interaction is displaced. When automation replaces interaction, AI undercuts SCT's dialogic engine, raising significant ethical, pedagogical, and practical concerns. These concerns warrant careful consideration before, during, and after integrating AI tools. When used without criticality, AI tools can disrupt or displace the social, dialogic nature of learning itself, replacing rather than mediating and extending interaction. This section highlights several key challenges: assessment; equity of access; ethics and bias; pedagogical concerns; and the preservation of human interaction—before turning to broader societal issues.

AI-informed summative assessment of writing presents many challenges. Teachers' use of automated essay scoring is somewhat controversial among students (Tossell et al., 2024), teachers (Wilson et al., 2021), and researchers alike (Goh et al., 2020; Murff, 2025). Already, large-scale testing companies have begun using AI scoring. However, critics argue that it can present an incomplete picture of students' understanding and writing skills, prioritizing arbitrary factors such as length (Goh et al., 2020), particularly when used without regard to student age and expected writing proficiency (Hannah et al., 2023). One researcher recommends using AI scoring as part of "validity-centered, human-AI collaborative systems that prioritize transparency and pedagogical alignment over computational supremacy" (Murff, 2025, p. iii). This inequity in scoring mirrors other AI concerns, such as funding and skill divides.

The digital divide in access to AI technologies exacerbates existing inequalities in education, employment, and civic participation, highlighting the urgent need for equitable solutions (Lahiri, 2024). While AI has already transformed the landscape of education and is actively shaping classrooms worldwide, millions of students do not benefit from these opportunities due to technological and socioeconomic barriers (Bulathwela et al., 2024). This algorithmic divide risks deepening disparities by amplifying discrimination and perpetuating systemic inequities (Yu, 2020). However, critical teachers can use AI to help identify more pathways to equity, particularly through adaptive platforms designed to meet the diverse needs of individual learners (Roshanaei et al., 2023). Bridging this divide requires a multipronged approach, including the development of inclusive AI tools; the use of open educational resources; and the implementation of transparent, human-centered algorithms (Bulathwela et al., 2024). Additionally, ensuring equal, ethical access to AI's benefits demands legal, ethical, and institutional strategies that promote fairness and inclusivity as artificial intelligence becomes increasingly prominent (Yu, 2020). Equity concerns shape the foundation of any responsible AI use in SLA.

These ethical issues reshape the social conditions of learning: mediation, participation, and representation. Regarding ethics, the challenges of AI implementation in classrooms include privacy and security issues, the potential amplification of social inequities and racial stereotyping, and power asymmetries (Weber, 2020). Conversations surrounding the ethical use of AI in K–12 classrooms also emphasize the importance of transparency, pedagogical appropriateness, and the protection of children's rights (Adams et al., 2021); such safeguards help maintain equitable conditions for interaction. Other research notes' concerns about negative perceptions, lack of technological skills among students and teachers, and issues with AI tool design (Crompton et al., 2022), each of which can constrain guided participation within the ZPD. To address these challenges, it is crucial to facilitate conversations about AI ethics with both teachers and students (Akgun & Greenhow, 2022) and to consider ethical principles specific to AI in education (Adams et al., 2021), thereby fostering inclusive and dialogic classroom communities.

AI systems often receive criticism for embedded biases, which can stem from the sources used to train the models themselves. When biased outputs enter language classrooms, students may feel less represented in learning materials, creating barriers to participation and learning. For example, these biases manifest in multiple ways, from reinforcing stereotypes (Weber, 2020) to excluding underrepresented voices (Zou & Schiebinger, 2018). For example, AI-generated stories may perpetuate cultural biases, limiting diverse representation in educational materials (Zheng & Stewart, 2024). Similarly, AI-generated images may reinforce racial and gender biases, perpetuating inequities in educational contexts (Makhortykh et al., 2021). These issues underscore the importance of scrutinizing AI tools and implementing safeguards to reduce bias in classroom applications (Selwyn, 2024). Without such measures, the risk of amplifying existing societal inequalities remains significant (Weber, 2020). Ethics and bias, therefore, intertwine, shaping how learners experience inclusion or exclusion in AI-mediated classrooms.

The adoption of AI in education raises a set of ethical and pedagogical questions, including authorship and ownership, trajectories of writing development, algorithmic bias, and the accuracy of generated content (Pack & Maloney, 2024). Additional concerns include the effects on learner autonomy, classroom ecology, and the distribution of pedagogical roles (Han et al., 2023). Although AI enables personalization and data-driven decision-making (Massala, 2023), it may also perpetuate inequality and disrupt pedagogical arrangements (Han et al., 2023; Selwyn, 2024; Weber, 2020). Framed within SCT, these dynamics reconfigure mediational means, participation structures, and the scaffolding through which learners progress toward independent performance; without critical design, such reconfigurations may narrow opportunities for dialogic interaction and equitable access to learning.

The reliance on AI tools in language learning also raises concerns about the development of critical thinking and creativity. Brown (2000) highlights the importance of students actively constructing meaning through cognitive engagement. When students rely heavily on AI for tasks such as writing or problem-solving, they may miss essential opportunities to reflect, analyze, and synthesize information (Alghamdy, 2023; Shalevska, 2023). Werse (2023) argues that the act of writing itself fosters critical thinking by requiring students to organize and evaluate their ideas, a process that AI may bypass. This automation of meaning-making removes opportunities for students to reflect, revise, and struggle productively within their ZPD, potentially short-circuiting the very processes that foster deep learning and undermining the accuracy and reliability of generated output (Pack & Maloney, 2024), especially in language and content domains in which the student writers are not experts. When AI provides incorrect or contextually inappropriate suggestions, it disrupts the scaffolding function it should serve, potentially misleading learners or reinforcing misconceptions. Together, these pedagogical concerns demonstrate that AI may inadvertently narrow rather than expand learning opportunities if not used critically.

In language learning, human interaction plays a central role in fostering authentic communication and SEL. While AI tools can provide immediate feedback and personalized instruction, they cannot replicate the relational dynamics that occur between teachers and students. Alghamdy (2023) notes that decentering the human element risks dehumanizing the learning experience, which can alienate students and diminish engagement. To mitigate this, educators can prioritize activities that maintain human connections and use AI as a complement rather than a replacement for interpersonal teaching. SCT emphasizes that learning is not just about input and output, but also about identity, interaction, and community. Overreliance on AI risks isolating learners from one another and from their teachers, weakening the social bonds and shared meaning-making that are essential to second language development.

Broader ethical concerns also arise at a societal level. These include the ecological costs of data-intensive AI implementations (Selwyn, 2024), as well as privacy and security issues, power asymmetries, equity in access to resources, and machine accountability (Weber, 2020). Reducing bias in AI use requires increased awareness among educators of pedagogical and ethical implications (Pack & Maloney, 2024). Researchers propose frameworks for critical AI literacy, such as integrating LLM tools to foster students' critical thinking (Joseph, 2023) and crafting stronger policies and regulations around AI use (Barnes & Hutson, 2024). There is a call for a more cautious approach to AI integration in education, emphasizing the need to navigate biases, ensure fairness, and foster inclusivity in AI-powered educational landscapes (Massala, 2023; Selwyn, 2024). As AI-enhanced applications become more prevalent in classrooms and administrative tasks, it is crucial to clarify and address ethical concerns to ensure that AI serves learners and educators effectively (Remian, 2019). However, eliminating bias may be challenging, as AI systems often learn from historical data that encodes past biases (Roselli et al., 2019). Ultimately, a cautious, reflective approach to AI integration that prioritizes teacher preparation and ethical considerations will be essential to harness its benefits without exacerbating educational inequalities or harming students' critical-thinking skills. This broader societal lens reinforces that critical AI literacy is not optional, but somewhat necessary, for equitable education.

6. Recommendations for teachers and researchers

Sustained, ethical adoption of AI in SLA requires iterative, community-based learning designs in which teachers critically mediate tools and researchers study equity-centered outcomes through an SCT frame. By synthesizing insights from existing principles and frameworks, this chapter provides educators and researchers with a comprehensive view of the possibilities and implications of AI for SLA. Regardless of the context, integrating AI into SLA requires a thoughtful and gradual approach to maximize its benefits and address potential challenges.

Integrating AI into SLA instruction requires thoughtful, strategic planning. Teachers should start small by implementing AI in one area, such as formative assessment, to gain familiarity and evaluate its impact before expanding to other domains. For instance, adaptive platforms like Smart English have demonstrated the potential to create low-stress, real-time assessments tailored to student proficiency levels (Hwang et al., 2022). A critical focus on equity is essential to ensure all students can benefit from AI tools, including those without reliable Internet access. In-class opportunities or offline alternatives can help bridge this digital divide and promote inclusivity (Bulathwela et al., 2024; Lahiri, 2024). Teachers must also critically evaluate AI tools, considering their validity, reliability, and appropriateness for SLA contexts (Han et al., 2023). Doing so ensures AI tools complement instruction, not replace it.

Teachers can also teach the critical use of AI by guiding students to evaluate AI-generated content carefully. By fostering digital literacy, teachers can help students discern when to trust AI suggestions and when to rely on their own understanding, a skill that promotes independent thinking and ethical engagement with technology (Joseph, 2023). Teachers can enrich the learning process by incorporating AI tools into collaborative activities (e.g., real-world simulations or scaffolded writing exercises) that encourage students to engage deeply with the target language. Finally, educators should reflect on their AI practices, both individually and within PLCs, to ensure these tools continue to meet the needs of their diverse learners in authentic, human-centered ways (Han et al., 2023). This reflective approach enables teachers to integrate AI responsibly, prioritizing relationships and creativity in language learning.

To realize AI's potential more fully in second language classrooms, teachers can draw on frameworks and support systems alongside tools and strategies as they evaluate and adapt their practice. PLCs and frameworks, such as technological pedagogical, content, knowledge (TPACK; Koehler & Mishra, 2009), can be beneficial as educators learn to integrate AI in ways that align with both their content goals and pedagogical values. As Hanshaw et al. (2022) note, these supports foster intentional technology use rooted in collaboration, reflection, and ongoing learning, rather than passive adoption. In this way, teachers become not only users of AI but also critical agents shaping its role in the language classroom.

School leaders and administrators play a crucial role in supporting the integration of AI into SLA instruction by providing teachers with the necessary resources and professional development. Administrators should explore teacher-training programs that include both technical and pedagogical aspects of AI integration. This training should address practical applications (e.g., utilizing AI for adaptive assessments) and critical issues, including data privacy and algorithmic bias (Han et al., 2023). Equally important is ensuring equitable access to AI tools across the school community. Administrators can achieve this by investing in inclusive technologies, fostering partnerships to provide necessary infrastructure, and allocating funds to address access gaps (Bulathwela et al., 2024). By creating an institutional culture that values the ethical use of AI, administrators can empower teachers to harness its potential for enhancing SLA while safeguarding against its risks.

Future research on AI in SLA should focus on both the practical applications of AI tools and their broader implications for equity and ethics. Researchers should empirically study the effectiveness of specific AI applications, such as adaptive learning platforms, in improving language proficiency across diverse student populations (Hwang et al., 2022). Additionally, there is a need to explore how students and teachers perceive the use of AI in classrooms, as these perceptions can significantly impact the adoption and success of such tools (Hannah et al., 2023). Beyond efficacy, researchers must investigate the ethical considerations of AI use in SLA, including potential biases, data privacy concerns, and the long-term impact on teaching practices (Pack & Maloney, 2024). By addressing these questions, researchers can provide evidence-based insights to guide educators and policymakers in the responsible integration of AI into language-learning contexts.

7. Looking ahead at SLA in the age of AI

AI is reshaping SLA not simply by introducing new tools but by reconfiguring how mediation, scaffolding, and interaction occur within learners' ZPD. Grounded in SCT, the promise of AI lies in its ability to extend participation, personalize pathways, and scaffold engagement, but only when used as a complement to, rather than a replacement for, authentic interaction.

Across planning, implementation, assessment, and community contexts, AI demonstrates potential to democratize access and enrich opportunities for multilingual learners. At the same time, it raises enduring concerns about equity, bias, and the risks of automation displacing dialogue. These tensions suggest that AI's role in SLA cannot be reduced to a catalog of tools; rather, it must be understood as a shifting terrain of opportunities and risks that require ongoing critical reflection.

Teachers remain central to this landscape. It is the human educator who creates classroom culture, orchestrates dialogue, and sustains relational learning. AI can provide scaffolds, such as sentence stems, adaptive tasks, and low-stakes practice; but teachers ensure that these supports lead to growth rather than dependency. In this sense, AI reframes teacher expertise—not diminished but deepened—as teachers decide when, how, and why to integrate AI into dialogic learning.

Ultimately, the integration of AI into SLA is less about technological adoption and more about conceptual evolution. As a field, SLA educators must grapple with how AI reshapes mediation, participation, and equity, and what it means for teaching to remain relational and human-centered in increasingly AI-saturated environments. By anchoring AI integration in SCT and related frameworks, educators and researchers can ensure that these tools advance the social, cultural, and ethical dimensions of language learning, rather than eroding them.

References

Adams, C., Pente, P., Lemermeyer, G., & Rockwell, G. (2021, June). Artificial intelligence ethics guidelines for K–12 education: A review of the global landscape. In I. Roll, D. McNamara, S. Sosnovsky, R. Lucking, & V. Dimitrova (Eds.), *Artificial intelligence in education* (pp. 24-28). Springer International. https://doi.org/10.1007/978-3-030-78270-2_4

Akgun, S., & Greenhow, C. (2022). Artificial intelligence in education: Addressing ethical challenges in K–12 settings. In J. MacIntyre & L. Medsker (Eds.), *AI ethics* (Vol. 2, pp. 431–440). https://doi.org/10.1007/s43681-021-00096-7

Alghamdy, R. Z. (2023). Pedagogical and ethical implications of artificial intelligence in the EFL context: A review study. *English Language Teaching, 16*(10), 87-98. https://doi.org/10.5539/elt.v16n10p87

Baber, H., Nair, K., Gupta, R., & Gurjar, K. (2023). The beginning of ChatGPT—A systematic and bibliometric review of the literature. *Information and Learning Sciences, 125*(7/8), 587–614. https://doi.org/10.1108/ILS-04-2023-0035

Barnes, E., & Hutson, J. (2024). Navigating the ethical terrain of AI in higher education: Strategies for mitigating bias and promoting fairness. *Forum for Education Studies, 2*(2), 1229-1246. https://doi.org/10.59400/fes.v2i2.1229

Bonner, E., Lege, R., & Frazier, E. (2023). Large language model-based artificial intelligence in the language classroom: Practical ideas for teaching. *Teaching English with Technology, 2023*(1), 23-41. https://doi.org/10.56297/BKAM1691/WIEO1749

Brown, H. D. (2000). *Principles of language learning and teaching* (4th ed.). Longman.

Brown, T., Mann, B., Ryder, N., Subbiah, M., Kaplan, J. D., Dhariwal, P., Neelakantan, A., Shyam, P., Sastry, G., Askell, A., Agarwal, S., Herbert-Voss, A., Krueger, G., Henighan, T., Child, R., Ramesh, A., Ziegler, D., Wu, J., Winter, C., ... Amodei, D. (2020). Language models are few-shot learners. In H. Larochelle, M. Ranzato, R. Hadsell, M. F. Balcan, & H. Lin (Eds.), *Advances in neural information processing systems* (Vol. 33, pp. 1877–1901). Curran Associates. https://proceedings.neurips.cc/paper_files/paper/2020/file/1457c0d6bfcb4967418bfb8ac142f64a-Paper.pdf

Bulathwela, S., Pérez-Ortiz, M., Holloway, C., Cukurova, M., & Shawe-Taylor, J. (2024). Artificial intelligence alone will not democratise education: On educational inequality, techno-solutionism and inclusive tools. *Sustainability, 16*(2), 781. https://doi.org/10.3390/su16020781

Choi, W., Park, H., & Chan, V. (2022). "You don't actually experience it until you're actually in it:" A translanguaging simulation for preservice teachers. *TESOL Journal, 13*(4). http://doi.org/10.1002/tesj.680

Crompton, H., Jones, M. V., & Burke, D. (2022). Affordances and challenges of artificial intelligence in K-12 education: a systematic review. *Journal of Research on Technology in Education, 56*(3), 248–268. https://doi.org/10.1080/15391523.2022.2121344

Cummins, J. (1979). Cognitive/academic language proficiency, linguistic interdependence, the optimum age question, and some other matters. *Working Papers on Bilingualism, 19,* 121–129. https://files.eric.ed.gov/fulltext/ED184334.pdf

DelliCarpini, M. (2009). Dialogues across disciplines: Preparing English-as-a-Second-Language teachers for interdisciplinary collaboration. *Current Issues in Education, 11*(2), 1–18. https://cie.asu.edu/ojs/index.php/cieatasu/article/view/1573/618

Dignum, V. (2021). The role and challenges of education for responsible AI. *London Review of Education, 19*(1), 1–11. https://doi.org/10.14324/LRE.19.1.01

Duca, M. (2023). Artificial intelligence. Myth and reality? *Annals of Philosophy, Social and Human Disciplines* (Vol. 4), 71-84. https://ibn.idsi.md/sites/default/files/imag_file/71-84_5.pdf

Echevarria, J., Vogt, M. E., & Short, D. J. (2008). *Making content comprehensible for English learners: The SIOP model* (3rd ed.). Allyn & Bacon.

Fathimatuzahro, M., & Rizkiyah, F. (2025). AI-driven feedback in EFL students' writing: Does it enhance their skills? *Celtic: A Journal of Culture, English Language Teaching, Literature and Linguistics, 12*(1), 637–657. https://doi.org/10.22219/celtic.v12i1.40594

Fu, M., & Li, S. (2022). The effects of immediate and delayed corrective feedback on L2 development. *Studies in Second Language Acquisition, 44*(1), 2-34. https://doi.org/10.1017/S0272263120000388

Garland, C. (2015). *Gamification and implications for second language education: A meta analysis* [Master's thesis, Saint Cloud University]. https://repository.stcloudstate.edu/engl_etds/40

Goh, T., Sun, H., & Yang, B. (2020). Microfeatures influencing writing quality: The case of Chinese students' SAT essays. *Computer Assisted Language Learning, 33*(4), 455–481. https://doi.org/10.1080/09588221.2019.1572017

Greene Nolan, H., Coenraad, M., & Young, V. M. (2024). Teaching partner, grading assistant, substitute teacher: Three ways teachers positioned an artificial intelligence tool in writing instruction. *Digital Promise.* https://doi.org/10.51388/20.500.12265/226

Hadi, M. U., Al Tashi, Q., Qureshi, R., Shah, A., Muneer, A., Irfan, M., Zafar, A., Shaikh, M. B., Akhtar, N., Wu, J., & Mirjalili, S. (2023). A survey on large language models: Applications, challenges, limitations, and practical usage. *Tech.Rxiv.* https://doi.org/10.36227/techrxiv.23589741.v1

Hall, T. (2019). SIOP helps, but it is not enough. *NYS TESOL Journal,* 31–36. https://doi.org/10.13140/RG.2.2.15611.00804

Han, B., Nawaz, S., Buchanan, G., & McKay, D. (2023). Ethical and pedagogical impacts of AI in education. In N. Wang, G. Rebolledo-Mendez, N. Matsuda, O. C. Santos, & V. Dimitrova (Eds.), *Artificial intelligence in education* (Vol. 13916, pp. 667–673). Springer Nature Switzerland. https://doi.org/10.1007/978-3-031-36272-9_54

Hannah, L., Jang, E. E., Shah, M., & Gupta, V. (2023). Validity arguments for automated essay scoring of young students' writing traits. *Language Assessment Quarterly, 20*(4-5), 399–420. https://doi.org/10.1080/15434303.2023.2288253

Hanshaw, J., Talbert, S., & Smith, J. (2022). Technology integration in the post-pandemic secondary classroom. In P. Bawa (Ed.), *Preparing faculty for technology dependency in the post-COVID-19 era* (pp. 195-211). IGI Global. https://doi.org/10.4018/978-1-7998-9235-9.ch011

Harper, C. A., & de Jong, E. J. (2009). English language teacher expertise: The elephant in the room. *Language and Education, 23*(2), 137–151. https://doi.org/10.1080/09500780802152788

Harry, A. (2023). Role of AI in education. *INJURITY: Journal of Interdisciplinary Studies, 2*(3), 260–268. https://doi.org/10.58631/injurity.v2i3.52

Hopfenbeck, T. N., Zhang, Z., Sun, S. Z., Robertson, P., & McGrane, J. A. (2023). Challenges and opportunities for classroom-based formative assessment and AI: A perspective article. *Frontiers in Education, 8,* 1–9. https://doi.org/10.3389/feduc.2023.1270700

Hwang, W. Y., Guo, B. C., Hoang, A., Chang, C. C., & Wu, N. T. (2022). Facilitating authentic contextual EFL speaking and conversation with smart mechanisms and investigating its influence on learning achievements. *Computer Assisted Language Learning 37(7),* 1632-1658. https://doi.org/10.1080/09588221.2022.2095406

Hyland, K. (2022). Second language writing instruction. In H. Mohebbi & C. Coombe (Eds.), *Research questions in language education and applied linguistics: A reference guide* (pp. 129-133). Springer Texts in Education. https://doi.org/10.1007/978-3-030-79143-8_24

Joseph, S. (2023). Large language model-based tools in language teaching to develop critical thinking and sustainable cognitive structures. *Rupkatha Journal on Interdisciplinary Studies in Humanities, 15*(4). https://doi.org/10.21659/rupkatha.v15n4.13

Kalaja, P. (1995). Student beliefs (or metacognitive knowledge) about SLA reconsidered. *International Journal of Applied Linguistics, 5*(2), 191-204. https://doi.org/10.1111/j.1473-4192.1995.tb00080.x

Kalaja, P., Barcelos, A. M. F., & Aro, M. (2017). Revisiting research on L2 learner beliefs: Looking back and looking forward. In P. Garrett & J. M. Cots (Eds.), *The Routledge handbook of language awareness* (pp. 222–237). Routledge.

Krashen, S. (1982). *Principles and practice in second language acquisition.* Pergamon. https://www.sdkrashen.com/content/books/principles_and_practice.pdf

Lahiri, A. (2024). Sociological implications of the digital divide: Exploring access to information and social inequality in the age of artificial intelligence and automation. I*nternational Journal of Multidisciplinary Approach & Studies, 9*(1), 156–167. https://doi.org/10.31305/rrijm.2024.v09.n01.019

Lampou, R. (2023). The integration of artificial intelligence in education: Opportunities and challenges. *Review of Artificial Intelligence in Education, 4*(00), e015–e015. https://doi.org/10.37497/rev.artif.intell.educ.v4i00.15

LeCun, Y., Bengio, Y., & Hinton, G. (2015). Deep learning. *Nature, 521*(7553), 436–444. https://doi.org/10.1038/nature14539

Lightbown, P. M. (2014). Making the minutes count in L2 teaching. *Language Awareness, 23*(1–2), 3–23. https://doi.org/10.1080/09658416.2013.863903

Long, M. H. (1983). Does second language instruction make a difference? A review of research. *TESOL Quarterly, 17*(3), 359–382. https://doi.org/10.2307/3586253

Makhortykh, M., Urman, A., & Ulloa, R. (2021). Detecting race and gender bias in visual representation of AI on web search engines. In Boratto, L., Faralli, S., Marras, M., & Stilo, G. (Eds.), *Advances in Bias and Fairness in Information Retrieval: Vol. 1418. Communications in Computer and Information Science* (pp. 36–50). Springer. https://doi.org/10.1007/978-3-030-78818-6_5

Massala, K. (2023). Navigating bias and ensuring fairness: Equity unveiled in the AI-powered educational landscape. *Apprendre et Enseigner Aujourd'hui, 13*(1), 37–41. https://doi.org/10.7202/1107542ar

McCallum, L. (2024). New takes on developing intercultural communicative competence: Using AI tools in telecollaboration task design and task completion. *Journal for Multicultural Education, 18*(1/2), 153-172. https://doi.org/10.1108/JME-06-2023-0043

McCarthy, J., Minsky, M. L., Rochester, A. S., & Shannon, C. E. (1956). A proposal for the Dartmouth summer research project on artificial intelligence. *AI Magazine, 27*(4), 12–14. https://doi.org/10.1609/aimag.v27i4.1904

Koehler, M., & Mishra, P. (2009). What is technological pedagogical content knowledge (TPACK)? *Contemporary Issues in Technology and Teacher Education, 9*(1), 60–70.

Mitchell, T. M. (1997). *Machine learning.* McGraw-Hill.

Mirzaei, A., & Eslami, Z. R. (2015). ZPD-activated languaging and collaborative L2 writing. *Educational Psychology, 35*(1), 5–25.

Murff, M. J. (2025). *AI-enabled innovations in automated essay scoring and feedback systems.* [Doctoral dissertation, Brigham Young University]. ScholarsArchive. https://scholarsarchive.byu.edu/etd/10933/

Nair, G. K. S., Abdul Rahim, R., Mohd Adam, A. F. b., Setia, R., Husin, N., Sabapathy, E., Mohamad, R., Mat So'od, S. M., Md Yusoff, N. I., Mohd Razlan, R., & Abd Jalil, N. A. (2012). Group work in the secondary ESL classroom. *Asian Social Science, 8*(13), 3–7. https://doi.org/10.5539/ass.v8n10p3

Nathan, L. S., & Hashim, H. (2023). A systematic review on gamified learning for improving language skills in ESL classrooms. I*nternational Journal of Academic Research in Business and Social Sciences, 13*(4), 481–493. https://doi.org/10.6007/IJARBSS/v13-i4/16873

Nation, I. S. P. (2001). *Learning vocabulary in another language.* Cambridge University Press. https://www.cambridge.org/core/books/learning-vocabulary-in-another-language/491314AA1B451AD04F3536000F1C9F0D

National Center for Education Statistics [NCES]. (2024). *English learners in public schools.* U.S. Department of Education, Institute of Education Sciences. https://nces.ed.gov/programs/coe/indicator/cgf

Newell, A., & Simon, H. A. (1976). *Computer science as empirical inquiry: Symbols and search. Communications of the ACM, 19*(3), 113-126. https://doi.org/10.1145/360230.360235

North Carolina Department of Public Instruction [NCDPI]. (2024, January 16). *North Carolina generative AI implementation recommendations and considerations for PK-13 public schools.* https://go.ncdpi.gov/AI_Guidelines

Pack, A., & Maloney, J. (2024). Using artificial intelligence in TESOL: Some ethical and pedagogical considerations. *TESOL Quarterly, 58*(2), 1007–1018. https://doi.org/10.1002/tesq.3320

Parks, P., Riches, C., & Hardin, K. (2023). Attracting and retaining English second language teachers. *Apprendre et enseigner aujourd'hui: Revue du Conseil pédagogique interdisciplinaire du Québec, 12*(2), 29–31. https://doi.org/10.7202/1101209ar

Pemberton, I. (2024). *Usage-based second language instruction: A context-driven multimedia learning approach.* Palgrave Macmillan.

Plonsky, L., & Sudina, E. (2021). Second-language strategy instruction. *Research questions in language education and applied linguistics: A reference guide.* Springer. https://doi.org/10.1007/978-3-030-79143-8_45

Raiaan, M., Mukta, S., Fatema, K., Fahad, N., Sakib, S., Mim, M. M. J., Ahmad, J., Ali, M. E., & Azam, S. (2024). A review on large language models: Architectures, applications, taxonomies, open issues and challenges. *IEEE Access.* https://doi.org/10.36227/techrxiv.24171183

Remian, D. (2019). *Augmenting education: Ethical considerations for incorporating artificial intelligence in education* [Master's thesis, University of Massachusetts]. https://scholarworks.umb.edu/instruction_capstone/52

Roselli, D., Matthews, J., & Talagala, N. (2019). Managing bias in AI. *Companion Proceedings of the 2019 World Wide Web Conference,* 539–544. https://doi.org/10.1145/3308560.3317590

Roshanaei, M., Olivares, H., & Lopez, R. R. (2023). Harnessing AI to foster equity in education: Opportunities, challenges, and emerging strategies. *Journal of Intelligent Learning Systems and Applications, 15*(04), 123–143. https://doi.org/10.4236/jilsa.2023.154009

Rumelhart, D. E., Hinton, G. E., & Williams, R. J. (1986). Learning representations by back-propagating errors. *Nature, 323*(6088), 533–536. https://doi.org/10.1038/323533a0

Schneider, M., Huss-Lederman, S., & Sherlock, W. (2012). Charting new waters: Collaborating for school improvement in U.S. high schools. *TESOL Journal, 3*(3), 373–401. https://doi.org/10.1002/TESJ.26

Selwyn, N. (2024). On the limits of artificial intelligence (AI) in education. *Nordisk Tidsskrift for Pedagogikk Og Kritikk, 10*(1), 3-14. https://doi.org/10.23865/ntpk.v10.6062

Setyaningsih, E., Asrori, M., Ngadiso, Sumardi, Zainnuri, H., & Hariyanti, Y. (2024). Exploring high school EFL teachers' experiences with Magic School AI in lesson planning: Benefits and insights. *Voices of English Language Education Society, 8*(3), 685–700. https://doi.org/10.29408/veles.v8i3.27700

Shalevska, E. (2023). English teaching just got an upgrade: How AI is revolutionizing TESOL. *International Journal of Education Teacher, 25*, 109–113. https://doi.org/10.20544/teacher.25.11

Sidek, H. M. (2011). ZPD, scaffolding, and syntax development. *International Journal of Humanities and Social Science, 1*(6), 23–34.

Simard, D. (2004). Using diaries to promote metalinguistic reflection among elementary school students. *Language Awareness, 13*(1), 34–48. https://doi-org.libproxy.bellarmine.edu/10.1080/09658410408667084

Tan, K., Pang, T., Fan, C., & Yu, S. (2023). *Towards applying powerful large AI models in classroom teaching: Opportunities, challenges and prospects* (Version 2). ARXIV Cornell University. https://doi.org/10.48550/ARXIV.2305.03433

Tossell, C. C., Tenhundfeld, N. L., Momen, A., Cooley, K., & de Visser, E. J. (2024). Student perceptions of ChatGPT use in a college essay assignment: Implications for learning, grading, and trust in artificial intelligence. *IEEE Transactions on Learning Technologies.* https://doi.org/10.1109/TLT.2024.3355015

Vygotsky, L. S. (1978). *Mind in society: The development of higher psychological processes.* Harvard University Press.

Weber, A. (2020). Ethics concerns in artificial intelligence use in education. In *INTED2020 Proceedings* (pp. 4539–4544). IATED. https://doi.org/10.21125/inted.2020.1262

Werse, N. R. (2023). What will be lost? Critical reflections on ChatGPT, artificial intelligence, and the value of writing instruction. *Double Helix, 11*(1), 1-4. https://doi.org/10.37514/DBH-J.2023.11.1.07

Wilson, J., Ahrendt, C., Fudge, E. A., Raiche, A., Beard, G., & MacArthur, C. (2021). Elementary teachers' perceptions of automated feedback and automated scoring: Transforming the teaching and learning of writing using automated writing evaluation. *Computers & Education, 168*, 104208. https://doi.org/10.1016/j.compedu.2021.104208

Yi, D. (2021). The effects of immediate versus delayed teacher feedback on L2 writing. *Studies in English Language Teaching, 9*(3), 69–78. https://doi.org/10.22158/selt.v9n3p69

Younas, A., Subramanian, K. P., Al-Haziazi, M., Hussainy, S. S., & Al Kindi, A. N. S. (2023). A review on implementation of artificial intelligence in education. *International Journal of Research and Innovation in Social Science, 7*(8), 1092–1100. https://doi.org/0.47772/IJRISS.2023.7886

Yu, P. K. (2020). The algorithmic divide and equality in the age of artificial intelligence. *Florida Law Review*, 72, 331–356. https://heinonline.org/HOL/LandingPage?handle=hein.journals/uflr72&div=11&id=&page=

Zheng, Y. D., & Stewart, N. (2024). Improving EFL students' cultural awareness: Reframing moral dilemmatic stories with ChatGPT. *Computers and Education: Artificial Intelligence, 6*, 100223. https://doi.org/10.1016/j.caeai.2024.100223

Zike, D. (2008). *Notebook foldables: For sprials, binders, and composition books.* Macmillan/McGraw-Hill.

Zou, J., & Schiebinger, L. (2018). AI can be sexist and racist—It's time to make it fair. *Nature*, 559, 323–326. https://doi.org/10.1038/d41586-018-05707-8

Chapter 2

Teacher in the loop: critically evaluating AI's role in language teacher preparation

Vera Sotirovska

Seattle University

Abstract

Now, more than ever, developing critical literacy skills for evaluating the use of generative artificial intelligence (GenAI) in classroom teaching has become essential in K–12 language education. As GenAI tools increasingly reshape classroom practice and policy discourse, educators benefit from cultivating critical AI literacy, a stance that extends beyond technical proficiency to encompass ethical, social, and environmental awareness. Grounded in established frameworks such as Lewison et al.'s (2002) dimensions of critical literacy and Muhammad's (2020) concept of Criticality, the chapter offers practical strategies for pre-service and in-service teachers. It illustrates how educators can evaluate, adapt, and address GenAI-generated content to support meaningful and contextualized learning experiences. Through sample lessons, reflective activities, and classroom discourse analysis, teachers are positioned as critical leaders in shaping how GenAI is implemented within language classrooms. Incorporating a TPACK approach, this chapter also addresses pressing educational concerns, including equity, bias, and sustainability, urging educators to develop a strong pedagogical voice and a sense of responsibility amid today's rapidly evolving digital landscape.

Keywords: Critical AI literacy; Teacher-in-the-loop approach; Teacher education; Critical literacy framework; GenAI tools

* * *

1. Introduction

Now, more than ever, critical literacy skills in utilizing artificial intelligence (AI) technologies for classroom teaching and instruction have become essential in K-12 language education. As AI legislation and curricular programming emerging globally, this urgency is heightened, prompting K-12 pre-service and in-service teachers to develop AI literacy skills and incorporate GenAI tools in their teaching and lesson planning with an ethic of care (Long & Magerko, 2020) and criticality (George et al., 2023). Building on this growing demand, this chapter problematizes the application of generative artificial intelligence (GenAI) in teacher education, focusing specifically on GenAI-produced content, rather than AI broadly, which is a field of computer science dedicated to building systems that perform tasks requiring human intelligence (Peñalvo & Ingelmo, 2023). Both terms appear throughout the chapter where each is most appropriate.

Technology companies market GenAI tools as timesaving thought partners, thereby positioning them as pedagogical and workflow enhancements, often in the form of chatbots, such as the GenAI pre-trained chat known as ChatGPT (Lo, 2023). As part of this marketing narrative, the creators of these tools promise teachers increased efficacy, greater capacity to tailor instruction, and support for data-informed instructional planning, even suggesting that GenAI can serve as a collaborative thought partner. Building on such promises, schools, administrators, and tech leaders increasingly mandate teachers become literate in GenAI tools and frameworks for their use, presenting these technologies as pathways to reduced workloads, rapid content creation, streamlined daily lesson planning, and increased student engagement.

With GenAI's increased ubiquity among students, who have become accustomed to using GenAI (e.g., the introduction of ChatGPT in late 2022), organizations at all levels of K-12 education have been developing innovative ways to both counter overreliance on these tools and harness them to imagine a more technologically conversant educational future. However, educators remain uncertain about how extensively GenAI should be used in the classroom, especially when they may lack a critical lens toward its implementation.

This chapter outlines pedagogical activities that foster a critically situated understanding of concepts, constructs, and propositions within the English language development (ELD) standards framework. The chapter concludes with practical implications, emphasizing holistic and ecocritical considerations regarding the potential environmental impacts of GenAI use and overuse.

1.1. What is AI literacy?

To understand the challenges surrounding GenAI in education, it is first important to examine how AI literacy is currently defined and conceptualized. Long and Magerko (2020) provide a working definition of AI literacy as a skill set that prepares individuals to critically evaluate and engage with AI technologies in various academic, professional, and social settings. Based on the premise that generative artificial intelligence (GenAI) simulates human-like productive and receptive skills (García-Peñalvo & Vázquez-Ingelmo, 2023), Leander and Burriss (2020) reimagine GenAI as 'an assemblage' that can endlessly reconfigure content with each new search and input. While there is an underlying need to evaluate GenAI outputs critically, there is a considerable lack of outlined ways for how teacher education programs and K-12 schools can engage responsibly in this practice. Given that different AI literacy frameworks describe critical engagement with AI as a key dimension, what this critical engagement actually looks like in practice remains unclear.

To build on this conceptual foundation, users must critically engage with GenAI and develop a deeper understanding of its mechanism for generating outputs, grounding this understanding in their prior knowledge and socio-political-historical contexts. Hence, this chapter begins by outlining a key difference between critical thinking and what Muhammad (2020) refers to as Criticality with a capital C, conceptualized as a literacy pursuit aimed at examining issues of power and promoting action for social justice. This chapter engages both critical thinking and Criticality as points of departure for addressing GenAI implementation in teacher education. For example, Dilek et al. (2025) propose a critical co-discovery approach to integrate AI literacy into teacher education by proposing a critical co-discovery approach that facilitates, aiming to foster "educators' reflective, critical, and participatory engagement with AI" (p. 294). Dilek et al.'s framework and other established frameworks in the field (Henriksen et al. 2025; Miao & Cukurova, 2024; Ng et al., 2021) identify ways for critical engagement with AI tools and outputs, which require explicit and scaffolded approaches for engaging with GenAI content.

GenAI tools are a common topic in education discourse and with the emerging AI for education organizations on the market (e.g., AI for Education, aiEDU, Code.org [the creators of Hour of AI], TeachAI, and Khan Academy), policy leaders such as UNESCO encourage states to produce legislation and guidelines for GenAI usage in educational settings. As a result, many school districts across the United States are beginning to implement procedures and draft policies for responsible and systematic incorporation of GenAI tools in classroom teaching. Despite this push for legislation in K-12 schools, teacher education programs have not been as quick to provide courses and supports for incoming teachers (Moorhouse et al., 2024), necessary for equipping teacher candidates with a skill set (Long & Magerko, 2020) to become critical educators and support student learning in a fast-changing landscape of GenAI technology.

1.2. What does a teacher-in-the-loop approach look like?

This chapter addresses the importance of developing critical AI literacy that operates from an ethical, sociopolitical, discourse-analytical, and contextually grounded stance. For this reason, the chapter highlights a teacher-in-the-loop approach in every stage of GenAI application, emphasizing teacher-evaluated outputs when interacting with chatbots, implemented as critical inquiry and problem-posing questions within a critical literacy framework (Lewison et al., 2002). In practice, the teacher-in-the-loop approach emphasizes teacher agency, justice-focused pedagogy, and responsible GenAI integration by positioning teachers as shapers of this technology rather than passive users. Munro (2021) introduced the human-in-the-loop (ML) concept, which inspired the term teacher-in-the-loop. Munro (2021) explains how in human-computer interaction, the computer learns from human feedback, emphasizing processes such as active learning, transfer learning, and ML, where humans control what information gets transferred to the machine. In this chapter, the term teacher-in-the-loop is used to foster teacher agency, Criticality, and positionality to reflect Munro's instructional equilibrium, balancing across teacher-led instruction, GenAI output, school GenAI usage policies, and student learning. Educators apply criticality and positionality to achieve Munro's instructional equilibrium by balancing teacher-led instruction, GenAI output, school GenAI usage policies, and student learning (Giroux, 1985).

2. Literature review

2.1. The intersections of technology-assisted language learning

In recent years, the fields of technology and education have become increasingly interconnected. This relationship is evident not only in classroom teaching practices—such as students effortlessly using Chromebooks alongside traditional paper materials—but also conceptually, in theory and research design. Technology not only informs educational practices but also shapes them. With a strong emphasis on multiliteracies, the notion that literacy extends beyond reading and writing to encompass various modes and modalities has gained traction (Kalantzis & Cope, 2023). This approach encourages students to engage with their full unitary repertoires (García & Kleifgen, 2020), positioning technological literacy as both a learning modality and a system in which students must become skilled. Consequently, students are encouraged to develop complex, multidimensional systems of literacy for effective, comprehensive, and engaged learning. Today's learners create knowledge within intricate systems of multiliteracies, modalities, and modes, with technology at the core of the meaning-making process (Kousa & Niemi, 2023).

The essential link between language, literacy, and technology has led to the emergence of teaching approaches such as Computer-Assisted Language Learning (CALL), where teachers act as facilitators of computer-mediated, supported, and enhanced language instruction, and Digitally Mediated Language Learning (DMLL), which incorporates digital technology into language teaching. Access to technology also implies access to language (De la Vall & Araya, 2023). In both methods, developing digital and technological literacy is crucial for students' engagement in language acquisition.

Additionally, we have seen a rapid adoption of Generative AI (GenAI) across all levels of education, along with research documenting educators' evolving beliefs about its role in language teaching and learning. Zhu and Wang (2025) conducted a systematic review of empirical studies on AI in language education (n=125). They found that topics such as writing supports, chatbots, translation tools, speech recognition, and intelligent tutoring systems were the most widely researched. Their findings indicated positive effects of automated writing evaluation systems and chatbots on students' language learning. However, they also noted a need for more qualitative studies that focus on less explored areas like reading, listening, grammar, and vocabulary (p. 26).

To enhance and expand on Zhu and Wang's findings, this chapter will explore additional seminal studies within the language teaching field that employ qualitative methods. For instance, Uwosomah and Dooly (2025) conducted a qualitative transnational study involving pre-service language teachers (n=46) from Spain and the United States. They collected data from three exit tickets in a technology-enhanced project-based language learning (TePBLL) course. Thematic analysis of pre-service teachers' engagement with GenAI revealed that they grappled with its complexities, recognized its potential for personalizing instruction, expressed concerns about its overuse, and made predictions regarding its future role in education.

Consistent with Moorhouse (2024), interviews with pre-service (n=10) and in-service teachers (n=17) as well as group discussions demonstrated varying degrees of understanding of GenAI, with both groups considering practical applications. In-service teachers specifically addressed opportunities (e.g., content generation for language teaching) and drawbacks (e.g., the risk of students becoming overly reliant on GenAI and ethical concerns about information accuracy and bias).

In light of these findings, scholars have emphasized the urgent need for critical classroom approaches that incorporate a teacher-in-the-loop strategy and scaffolded, hands-on examples to help educators navigate the challenges and innovations brought about by GenAI. To explore possibilities for integrating this technology into practice, Sperling et al. (2024) suggest reframing the conversation about GenAI to focus on the quality and substance of teacher instruction. They argue that there is currently a gap in educational strategies regarding how teachers can address GenAI in K-12 education, along with a lack of essential requirements for teacher educators to prepare future educators in this domain (p. 9).

2.2. Guiding conceptual GenAI frameworks in the field of language education

All the reviewed frameworks emphasize the importance of critical AI literacy as either a foundational guide or a necessary skill for preparing educators for the rapidly changing landscape of teacher education. Despite numerous studies highlighting the need for critical AI literacy, Sperling et al. (2024) argue in their scoping review that practice-oriented literature in the field often reflects "a somewhat uncritical promotion of AI" that relies on "the existence and use of digital artifacts" (p. 10). To address the need for a critical approach to AI literacy when evaluating outputs from generative AI (GenAI), educators should have opportunities for scaffolded lesson planning, focusing on accessibility, differentiation, and pedagogical learning. Specifically, pedagogical activities that foster a critically situated understanding of concepts, constructs, and propositions within the English language development (ELD) standards framework give educators the tools they need to be effective in the classroom. The chapter concludes with practical implications, emphasizing holistic and ecocritical considerations regarding the potential environmental impacts of GenAI use and overuse.

2.3. Emerging applications of GenAI tools in language teacher education

A significant number of teachers report using GenAI in their classrooms. According to data from the Rand Corporation and the Center on Reinventing Public Education, 18% of teachers have utilized GenAI and are seeking additional knowledge to integrate it effectively into their teaching practices (Diliberti et al., 2024). Complicating matters, both private and public institutions are navigating challenges related to AI legislation, particularly concerning K-12 school district policies across the country, and the extent to which AI should be incorporated into student learning.

Amid calls for reimagining K-12 education, recent developments in GenAI have led to the establishment of innovation hubs at large public universities. These hubs focus on developing tools and methods to assist teachers in incorporating AI technologies into K-12 classrooms (Belda-Medina & Calvo-Ferrer, 2022). For example, professors and community partners at

AmplifyLearn.AI, a research hub at the University of Washington, have initiated the development of a GenAI platform designed specifically for educators. Additionally, an independent GenAI platform called Colleague AI (2024) is being co-developed in partnership with teachers, students, and school district leaders. One of its most significant advantages is its ability to plan K-12 lessons and create activities and assessments while ensuring compliance with the Family Educational Rights and Privacy Act (FERPA).

Many leading educational institutions and colleges across the country are encouraging both pre-service and in-service teachers to engage with GenAI tools; they envision a pedagogical approach where these tools act as assistants, collaborators, and thought partners (Liu & Sun, 2023). For example, Colleague AI features two teacher assistant bots, Claire and Charlie. These AI assistants operate on conversational platforms like ChatGPT and can generate lesson plans, rubrics, and podcasts; troubleshoot instructional strategies; design and simulate instructional scenarios; and tailor content to support teachers in their instructional decision-making.

Chatbot Claire, as part of the Colleague AI platform, enhances instructional support by providing detailed explanations of its outputs, fostering critical thinking, and promoting reflective practice in educational settings. Claire provides outputs with two accompanying explanations: one focuses on the reasoning behind the output, called the "Thought Process," and the other, titled "Reflection," summarizes how well the output satisfies the user's prompt (University of Washington College of Education Newsroom, 2024). When I asked the chatbot Claire to explain the purpose and value of including the Thought Process and Reflection sections in its output, it explained that these sections aim to model "transparent reasoning, quality assurance, and educational modeling, which demonstrate critical thinking and reflective practice." It further emphasized the importance of this approach: "By showing this process, I'm modeling the kind of thoughtful analysis that's valuable in educational settings" (Colleague AI, 2024).

The use of AI chatbots like Colleague AI's Claire illustrates the potential for simulating critical thinking skills that educators need, but more research is necessary to understand how these technologies can effectively cultivate these skills in educational contexts. Clearly, Colleague AI (2025) offers tool that simulate critical thinking and promote reflective practice—the very qualities that scholars argue teacher candidates and educators need to effectively engage with GenAI outputs. However, further research is needed to explore how these GenAI chatbots help educators and students cultivate these skills within their own contexts, particularly in terms of the necessary technological know-how and the transactional qualities of GenAI-mediated pedagogical learning.

3. Current perceptions of GenAI in education: can GenAI tools be equitable?

AI-generated outputs are trained on Internet-wide historical data (Park, 2024); these outputs can include stereotypes, racial bias, and inaccuracies (Shieh et al., 2024). Such inaccuracies, called hallucinations, may convey distorted views of people and the world, perpetuating regimes of whiteness (Hurley, 2024). This phenomenon can propel students into a new era of mis- and disinformation (Williamson, 2024) where technocratic systems operating with deep-rooted biases may reinforce this effect. GenAI tools also exacerbate information appropriation. Additionally, widespread use of GenAI contributes to high energy consumption and significant carbon and water footprints, thereby raising awareness of environmental impacts (Bhaskar & Seth, 2024). Plans for green data centers, renewable energy, carbon offsetting, and global regulations are intended to assuage fears (e.g., EU AI Act; Doo et al., 2024), though such plans remain unrealized.

In addition to the significant environmental footprint and the cost of building and maintaining these large data centers, De Roock (2024) suggests that "GenAI technologies should be understood within the historical foundations of literacy even as they require a profound reconsideration of conventional notions of the productive value of literacy" (p. 591). De Roock questions the current framings of GenAI as an equitable information distributor and socioeconomic equalizer when teachers are asked to incorporate GenAI into their teaching with minimal or no preparation and are tasked with preparing K-12 students to do the same (Sperling et al., 2024).

Guided by these scholarly insights, I applied the critical literacy beliefs framework (Lewison et al., 2002; Sotirovska & Vaughn, 2022, 2023a, 2023b) to evaluate GenAI content for teacher preparation in K-12 language and literacy education. This framework breaks down critical literacy dimensions into guiding questions that help foster critical pedagogical dispositions in pre/in-service teachers. In the context of GenAI, teacher candidates develop critical literacy dispositions regarding GenAI applications across the four critical literacy dimensions. To achieve this, they examine what is commonplace in GenAI content (Lewison et al., 2002) and assess its relevance to their teaching contexts through critical evaluation, adaptation, and output modification (Kohnke & Zou, 2025). Teacher candidates also are encouraged to evaluate outputs for their accuracy in representing multiple perspectives (Lewison et al., 2002) and for how well they reflect the sociocultural contexts of their school communities. Alongside developing their technological, pedagogical, and content knowledge (Kohnke et al., 2023), teacher candidates and educators must also cultivate critical dispositions toward civic responsibility. Through problem-posing questions, the critical literacy framework invites pre/in-service teachers to collectively reimagine classrooms as egalitarian spaces, where both students and teachers develop skills and critical consciousness regarding GenAI (Sperling et al., 2024).

4. Significance language education

Critical AI literacy is an emerging pedagogical approach that educators need to develop, regardless of their stance on the GenAI continuum, whether they resist, embrace, or advocate for GenAI in language education. Toncelli and Kostka (2024) promote the need for more research in the field of language education, particularly "studies that explore how educators incorporate GenAI into their practice" (p. 78). An essential part of this research is understanding how GenAI outputs can highlight pedagogical possibilities in two key ways: (1) by focusing on students' complete unitary repertoires (Otheguy & García, 2024) and (2) by designing lessons that incorporate problem-posing pedagogy with culturally and linguistically responsive and sustaining orientations. This approach is crucial before we can explore how GenAI technologies can enhance language education for both teachers and students.

Corp and Revelle (2023) encourage the investigation of ChatGPT's potential as a lesson-planning tool, focusing on practices that enhance teacher candidates' skills and dispositions for evaluating, critiquing, and modifying outputs. The scaffolded lesson plan model generated with ChatGPT (Appendix A) demonstrates how a ChatGPT-generated lesson can foster critical engagement opportunities (Muhammad, 2020), as well as opportunities for modification and redefinition (Kohnke & Zou, 2025) that foster critical literacy dispositions (Sotirovska & Vaughn, 2023a, 2023b) in three essential areas of language lesson planning: (1) ELD proficiency standards, (2) culturally responsive and sustaining education (CRSE)-informed practices, and (3) opportunities for scaffolding. This practice-oriented approach bridges theory and practice while guiding teacher candidates in their instructional decision-making, all grounded in a commitment to responsible and critically minded implementation of generative AI technologies.

5. Conceptual framework

Critical literacy refers to the understanding that texts are constructed within socio-political, historical, and cultural contexts of power (Lewison et al., 2002); this concept is especially relevant for reading and interpreting content generated by generative AI (GenAI) and is particularly important in the context of language teaching. Building on Lewison et al.'s (2002) 20-year scoping review of critical literacy literature and their resulting framework—which includes four dimensions: (1) disrupting the commonplace, (2) exploring multiple perspectives, (3) focusing on sociopolitical issues, and (4) taking action for social justice—I developed critical probing questions (Sotirovska & Vaughn,

2023a, 2023b) to evaluate GenAI's potential in language teaching and learning. This framework emphasizes a human-centered design approach that prioritizes equity, responsibility, and safety, including a description of how GenAI models align with educational theories and evidence-based practices while adopting a teacher-in-the-loop approach.

5.1. Disrupting the commonplace

Disrupting the commonplace is an important aspect of critical literacy that helps educators analyze who is represented in texts and curricula; it empowers them to change, adapt, and reimagine these representations in the interest of equity and justice. The critical AI literacy framework is based on two key assumptions: (1) everything can be considered a text that can be interpreted (Freire & Macedo, 2005), and (2) AI-generated content should be critically analyzed and modified by teachers before being introduced into the classroom (Kohnke & Zou, 2025; Sperling et al., 2024).

Questions educators can ask to critically evaluate GenAI tools: disrupting the commonplace

Who and what does the output represent and privilege (Sotirovska, 2024; Sotirovska & Vaughn, 2023a, 2023b)?

How does the output engage, reflect, and account for the classroom context of learning?

How do outputs affirm or challenge what is already conveyed in the curriculum?

To what extend do you find GenAI "reading, listening, grammar, and vocabulary" (Zhu & Wang, 2025, p. 26) content applicable?

What supports these inferences?

5.2. Focusing on sociopolitical issues

Focusing on sociopolitical issues is a dimension that situates GenAI outputs within the current sociopolitical landscape and reflects a deeper awareness of intersectional identities, histories, and contexts.

Questions Educators Can Ask to Deepen Awareness of Intersectional Identities, Histories, and Contexts

How and to what extent do GenAI outputs showcase an awareness of diverse classroom contexts (e.g., reflective of multilingualism, representative of students' home and heritage languages and inclusive of diverse sociohistorical and geographical contexts)?

How and to what extent do GenAI outputs include multiple perspectives?

How and to what extent do GenAI outputs provide scaffolds for diverse learner profiles, facilitating inclusive learner experiences?

How and to what extent do GenAI outputs enrich the curricula and prompt teachers to consider multiple perspectives (e.g., layering texts [Muhammad, 2020], showcasing a topic from various angles and sociocultural contexts)?

5.3. Exploring multiple perspectives

Examining multiple perspectives enhances representation and encourages educators to reflect on the connection between perspectives and agency by questioning who is represented and who has the power to influence (Vaughn et al., 2020).

Questions Educators Can Ask to Determine Who is Platformed and/or Who Has Agency to Exert their Influence.

Does the GenAI tool provide outputs and templates where multiple perspectives are represented?

What steps need to be taken for this tool to yield outputs that will meet the criteria for exploring multiple perspectives?

Does the tool operate from the premise that knowledge is co-constructed and must represent diverse funds of knowledge, place-based approaches, and justice-oriented pedagogies?

5.4. Taking action to promote social justice

Taking action and promoting social justice is a dimension that encourages exploring ways can be leveraged as a tool for technology social justice-oriented learning. The two big promises of GenAI-assisted instruction largely focus on differentiation and inclusion, and on reducing the teachers' workload (Henriksen et al., 2025).

Questions Educators Can Ask to Take Action and Promote Social Justice

How do GenAI tools support educators in providing multiple means of representation, access, and expression of knowledge (Cope et al., 2021)?

What are the affordances, opportunities, and consequences of using GenAI in teacher and student learning?

What are the implications of habitual use of these tools (Alasadi & Baiz, 2023)?

In what ways, if any, can you envision taking action for social justice by leveraging these tools?

What are the implications for equity in and access to education?

Concerning sustainability, what is the environmental footprint of GenAI tools that are recommended for classroom use?

Is this information prominently posted on the GenAI tool's websites?

How can educators take action to inquire about the environmental footprint and consider sustainability when using and addressing GenAI tools in the classroom?

Currently, there are numerous claims that generative AI (GenAI) has the potential to promote equity and equality by enhancing access to education (Luckin et al., 2016). I have included a series of questions that educators can consider to critically examine these claims and engage in thoughtful inquiry within their practice.

Questions Educators Can Ask to Increase Equity and Equality

Does GenAI make language learning more accessible? If so, how?

How can GenAI tools support a teacher-in-the-loop approach in language teaching practice?

How are GenAI technologies relevant to the lives of the community of learners we are serving?

What transformative learning experiences do said technologies afford students? What might these experiences look like?

How is GenAI supporting students to become critically engaged language learners?

To what extent are these tools effective in reducing workload and enhancing language teachers' lesson-planning efficacy?

How have language teacher classroom experiences changed upon the implementation of these tools?

How have students' experiences changed upon the implementation of these tools?

6. Critically evaluating GenAI outputs and unpacking ELD standards

Scholars encourage teacher candidates to put themselves in their students' shoes (Guerrettaz et al., 2022) and plan lessons that consider students' full range of abilities (García, 2013) to effectively address their emotional and academic needs. More specifically, teacher educators are responsible for ensuring that language teacher preparation programs successfully incorporate content-rich multiliteracies into English language instruction (Sotirovska, 2025), especially when creating lessons for multilingual learners. Guided by the framework developed by Lewison et al. (2002), this chapter presents an alternative thought process that illustrates Criticality in reading and interpreting GenAI outputs in language teaching. It aims to critically examine the potential benefits of enhanced support, greater freedom, and improved educational outcomes for all through the use of technology (Hargrave et al., 2024).

6.1. Developing critical linguistic praxis: analysis of a GenAI lesson scaffold

Consider the following scenario in which teacher educators and pre-service teachers analyze GenAI content to support the specified English Language Development (ELD) standards under the category of Language for Social and Instructional Purposes. This category states that "English language learners communicate for social and instructional purposes within the school setting" (World-Class Instructional Design and Assessment English Language Development Standards Framework [WIDA ELDSF], 2020). In this scenario, they select a specific language function, such as "explain," and use a chatbot to generate responses related to content domains, infer language demands, and outline skill expectations. Given that developing linguistic practices takes years (Guerrettaz et al., 2022), technology can play a crucial role in analyzing the language complexity of classroom materials. This analysis includes (1) evaluating Lexile levels, (2) sentence complexity and structure, (3) tiered vocabulary, (4) syllabification, (5) morphological chunking, and (6) other examples of discourse-level language analysis. Because evaluating classroom materials is a nuanced process, it is essential to consider language demands and content from multiple perspectives to ensure that students have access to inclusive lessons. For this reason, instructional decision-making should follow a approach, where teachers adapt technology rather than technology dictating teaching methods.

Currently, the use of GenAI primarily focuses on generating content and planning instruction. I propose a shift in how these tools are utilized. More specifically, applications of GenAI in K-12 language teaching should emphasize collaborative instructional planning, using GenAI tools for content ideation (Kohnke et al., 2023), creating targeted language prompts (Ma et al., 2024), adapting teaching materials and scaffolds (Kim & Adlof, 2024), and generating prompts based on language proficiency standards through guided inquiry (Kostka & Toncelli, 2023). To be effectively prepared for today's classrooms, teacher candidates specializing in multilingual education must be ready to implement K-12 GenAI guidance frameworks. However, these frameworks are often too broad, requiring new teachers to translate them into practical strategies without adequate support for students' language development goals. To translate these frameworks into practice, teacher educators can guide candidates to start by unpacking the WIDA ELD Standards Framework for language expectations and uses and develop specific ELD standards statements. This action constitutes the first step in guiding teacher candidates to translate pedagogical frameworks into classroom practice.

To further clarify the WIDA ELD standards framework, the proficiency level descriptors (PLDs) should align with the ELD Standard Statement. This includes the key dimensions of language use and language expectations that have been identified as the focus of an instructional task. These elements describe how an individual student performs a task across the six levels of language proficiency. For example, teacher candidates and educators can select one of the grade clusters (K, 1, 2-3, 4-5, 6-8, and 9-12), choose a communicative mode—either interpretive (listening, reading, and viewing) or expressive (speaking, writing, and representing)—and identify one of the language dimensions (e.g., word/phrase for precision, sentence level for grammatical structure, or discourse level for organization, cohesion, and density of language) while considering the context, including topics, audiences, and situations (WIDA, 2020).

As an example, I am including the complete task prompt from which a lesson plan for multilingual language learners was generated, using the specified WIDA standard and lesson orientation as a guide:

ELD-LA.2-3.Narrate.Expressive: Multilingual learners construct language arts narratives that

- Orient the audience to context;
- Develop a story with sequences of time and events, complications, resolutions, or endings; and

- Engage and adjust for the audience (p. 16).

The purpose of this lesson plan was to engage teacher candidates in critical pedagogical learning, helping them develop the craft of lesson planning as well as skills in inductive GenAI learning and output analysis.

The following prompt was created to support the following ELD standard:

> Develop a lesson plan using the guidelines based on ELD-LA.2-3.Narrate.Expressive. The plan should incorporate culturally responsive and sustaining pedagogies suitable for a diverse, multilingual classroom, including example materials and resource suggestions.

In the ChatGPT-generated lesson plan (Appendix A), I include "Pause and Reflect" sections to reinforce the essential skills outlined in the teacher-in-the-loop approach, such as noticing, evaluating, and reflecting on AI-generated outputs for critical pedagogical learning. This example lesson can be modified in three key areas: (1) by integrating students' prior knowledge and funds of knowledge, (2) by incorporating translanguaging, and (3) by providing opportunities for student agency.

To effectively incorporate students' prior knowledge and translanguaging practices, the lesson can be adjusted to focus on stories from the students' families and community experiences. Students can interview their parents or guardians, which is a skill that needs to be modeled in the classroom (Souto-Manning & Martell, 2016). In the lesson simulation, teachers might start with a shared reading of community stories, including those from the students themselves, and link this read aloud to the lesson's learning objectives. Next, students and teachers can discuss who in the community they can interview, documenting their stories in a classroom almanac. This almanac serves as an artifact of collective memory, creating a living text of the classroom's literacies. Language supports may include sentence stems, examples of verb tenses for composing narrative structures (such as the suffix -ed), and personal stories as mentoring texts to help students identify key elements of storytelling in both reading and writing contexts.

In the teacher preparation classroom, teacher educators can demonstrate how to critically question GenAI outputs by starting with a think-aloud segment framed as a form of wondering. After analyzing the scaffolded GenAI lesson and engaging in reflection, teacher candidates develop positionality statements as a culminating assignment focused on critical literacy. This activity allows them to explore how GenAI tools contribute to or challenge concepts such as multiliteracies, translanguaging, and bias. To enhance teacher candidates' critical AI literacy in language teaching and learning, teacher educators can use the following questions as a guide (Appendix B) when conducting a collective analysis of the GenAI scaffolded lesson.

Using the questions in Appendix B, teacher educators can guide candidates in incorporating critical tenets to develop AI literacy. Sardinha's (2024), in their research, highlights the significant disparity between human and AI-generated content; these questions can be applied as evaluative criteria of GenAI output.

This chapter advocates for the strategic use of GenAI tools as supportive resources in pedagogical learning. These tools can help exemplify, visualize, and clarify content while encouraging teacher candidates to develop their own pedagogical voices, positionality, and practices through inquiry. This process involves layering texts (Muhammad, 2020) and exploring multiple perspectives (Lewison et al., 2002) within the socio-political and historical contexts in which they will teach. I advocate for the teacher-in-the-loop approach, as only teachers can provide the necessary pedagogical context, model agency, and incorporate critical thinking to examine, evaluate, or create language and literacy curricula; they also teach from a perspective of care that GenAI cannot replicate. Relying too heavily on GenAI may lead to a task-driven environment, resulting in inequitable learning outcomes rather than fostering open-ended inquiry, enriching literacy experiences, and promoting metacognitive practices.

7. Ways to strengthen practicums for pre-service teachers focusing on language education: reflective practices

Teacher candidates are encouraged to develop critical literacy practices when engaging with GenAI outputs. Specifically, when analyzing the lesson scaffold (see Appendix A) and content produced for teacher candidate learning through ChatGPT or any GenAI tool, candidates should receive guidance in the production of critical annotations of GenAI outputs. These annotations should reflect their teacher education program competencies as well as their field experience, as demonstrated in the scaffolded example. This chapter also includes recommended practices for fostering critical AI literacy among teacher candidates.

7.1. Reflective practices for deeper understandings of GenAI outputs

Reflective practice 1: Teacher candidates are encouraged to explore school district GenAI policy in their practicum placement schools to understand how GenAI is incorporated at their internship school, completing a brief reflection detailing their school-level policy guidelines regarding GenAI implementation.

Teacher candidates can conduct an inquiry analysis at their placement schools to understand the GenAI policies implemented at both the school and district levels. One example is MagicSchool, a FERPA-compliant GenAI tool that serves as an interactive platform for both teachers and students. It is currently being used in schools across the United States, boasting 6 million active accounts to date (MagicSchool, n.d.). MagicSchool was introduced to help reduce teacher workloads and make learning more accessible for students. This aligns with the principles of universal design for learning (UDL), a framework that promotes accessibility in teaching and learning through diverse multimodal approaches and inclusive practices (Dwyer-Kuntz & Kuntz, 2024).

MagicSchool's user resources include professional development on aligning learning objectives with tailored and differentiated assessments along with ancillary resources (e.g., rubrics, exemplars and non-exemplars, and multiple-choice questions). Tools to support MagicSchool users include text adaptation and scaffolding creation (e.g., text scaffolder, translator, and rewriter); these tools connect users to SMART goals and progress trackers. Highly praised supports are those aligned with Tier 2 and Tier 3 interventions in the multi-tiered system of support (MTSS) and Individualized Education Programs (IEPs).

Teacher educators can involve teacher candidates in usability testing of a GenAI tool, such as MagicSchool, to evaluate how effectively it supports student learning during their practicum experiences. This testing process includes examining the frequency of use, the purpose of the tool, and the outcomes for students when utilizing GenAI platforms for learning. Teacher candidates are encouraged to describe the tool and its intended purpose, outline its features, assess whether empirical evidence supports its effectiveness in K-12 education, and identify case studies and research that support its use in language learning.

Reflective Practice 2: Teacher candidates engage in discourse analysis of GenAI outputs to develop a discerning lens focusing on human and GenAI created content for developing their linguistic and pedagogical knowledge and critical analysis skills.

Discourse analysis of GenAI-generated outputs can be conducted at multiple levels. For example, teacher candidates at various stages of their degree programs can engage in paragraph analysis guided by the question: What do you notice about the language in GenAI outputs? The first step in analyzing GenAI outputs is to understand how a specific prompt generates a particular response. As previously stated in this chapter, linguistic content knowledge is critical for teaching and learning languages (Guerrettaz et al., 2022). Therefore, examining how these outputs are composed can promote understanding of the potential advantages and limitations of incorporating them into language teaching and learning. Teacher candidates can explore questions such as: What types of vocabulary words are frequently present in these outputs? Developing responses to this question may include conducting a tiered vocabulary analysis may involve looking for and categorizing Tier 1, 2, and 3 vocabularies. Additionally, semantic mapping can be employed to clarify the relationships between words and their cohesive ties; such activities benefit teacher candidates as they develop scaffolded texts to help students practice morphological, syntactic, and semantic vocabulary skills. Candidates can also evaluate the alignment between these outputs and the curriculum and topics they are teaching. It is important to consider whether the outputs are developmentally appropriate for the age groups of the students with whom they are working during their practicums.

Additional analysis could focus on the following areas of inquiry:

Questions Teacher Candidates Can Ask to Analyze Language Outputs

What levels of linguistic complexity do you notice in examining outputs (e.g., syntactic and morphological complexity by noticing sentence composition and frequency of monosyllabic and multisyllabic words)?

What words do you notice appearing most frequently across outputs?

What dominant phrasing and punctuation are common?

In other words, teacher candidates are invited to analyze the language so that they can begin to familiarize themselves with outputs and come to their own conclusions about what can and cannot be accomplished by using GenAI.

Questions Teacher Candidates Can Ask to Understand What Can and Cannot be Accomplished with AI

How does the generated output convey linguistic complexity and/or accessibility?

How does the output reflect hedging language, modal verbs and expressions, and frequency of specific parts of speech, imagery, and other relevant discourse markers?

What is missing?

Teacher candidates can heighten their awareness of the disparities between human and AI-generated outputs (Sardinha, 2024) and develop linguistic knowledge that goes beyond checking for output accuracy to understand how arguments are being constructed from a social, cultural, and linguistic point of departure.

Reflective practice 3: Teacher candidates engage in multiple levels of comprehension analysis of outputs may include the following:

Literal comprehension (e.g., who, what, where, and when): this level of comprehension is aimed at determining prompt and output alignment.

Questions Educators Can Ask to Engage in Multiple Levels of Comprehension Analysis of Outputs
What is the message?
How is the argument being built/or how is the topic discussed?
What timeframe does it span and to what spatial context does it apply?

Inferential comprehension: this level of comprehension is aimed at developing metacognitive practices by engaging teacher candidates' knowledge funds as a lens for interpreting GenAI outputs.

Questions Educators Can Ask to Develop Metacognitive Practices
How can I use this output and my own prior/funds of knowledge together to arrive at a deeper meaning or understanding of the topic?
What are the opportunities to apply this generated activity to my own context of teaching and learning and what contexts should I further explore?

Evaluative comprehension: this level of comprehension is aimed at evaluating the extent to which GenAI application is supporting, transforming, and advancing teacher candidates' content and pedagogical knowledge.

Questions Educators Can Ask to Evaluate the Extent to which GenAI Application is Supporting, Transforming, and Advocating Teacher Candidates' Content and Pedagogical Knowledge
To what extent is this information relevant to my teaching and learning?
What sources are referenced?
How does this output create opportunities for meaning making and productive struggle?
Is the output enjoyable, does it appeal to students' ethos, pathos, logos?

How does it create engagement and opportunities for student agency and joy (Muhammad, 2020)?

Reflective practice 4 (inquiry-focused models): Teacher candidates participate in Workshops 1 and 2; explore flipped classroom model.

The Workshop 1.0 (Kilbane & Milman, 2013) begins by introducing the lesson objectives, followed by a teacher educator modeling a mini lesson of GenAI output analysis. The teacher educator uses direct and explicit instruction to introduce specific critical AI literacy skills. For example, a teacher educator can model a think aloud about reading for GenAI bias in a generated paragraph for a particular grade level. During this mini lesson, teacher educators can explore whose perspective is represented in the output by highlighting the subject structures and the cohesive ties, structuring pause and reflect sections for teacher candidates to think deeply about the text. In the application stage, teacher candidates continue to work in small groups while the teacher educator facilitates and confers with candidates. Toward the end, the group is brought together to collectively reflect on and share inferences from the activity, identifying look-fors and take-aways that can apply to their practicums.

The Workshop 2.0 model (Kilbane & Milman, 2013) is structured around an inquiry-driven task; however, in this lesson model, the teacher candidates are first asked to look for evidence of GenAI bias by analyzing the output on their own. Teacher candidates grapple with the following question: how can we identify GenAI bias in this specific output by finding evidence of positionality? In pairs or small groups, teacher candidates participate in close reading and annotating as well as developing critical questions. Only after this inquiry stage is completed does the teacher educator provide direct and explicit instruction in the form of a mini lesson.

The flipped instruction lesson model (Kostka & Marshall, 2018) prioritizes deeper engagements with texts (including GenAI outputs) prior to class, where teacher candidates are guided by an inquiry approach. In this lesson model, teacher candidates engage with the actions and cognitive processes of Bloom's taxonomy in a reverse order. Before class, they focus on understanding, applying and analyzing assigned materials, while during class, they apply their knowledge, negotiate meaning in a group setting, form stances and plan for their enactment. The inquiry analysis concludes with a reflection.

The emphasis is on the productive struggle and learning by discovery through which Criticality emerges (Muhammad, 2020). Teacher candidates grapple with big ideas by investigating layered texts, analyzing outputs, reading and annotating case studies, forming emerging perspectives and applying them in immersive K-12 classroom experiences. The flipped classroom instruction model (Kostka, 2023) could also look like organizing candidate-led workshops on critical inquiry into GenAI outputs by exploring practitioner articles that bridge theory and practice (Kostka & Toncelli, 2023) in the language teaching context.

This chapter concludes with recommendations for incorporating select multimodal artifacts, including a debate on GenAI between Sal Khan, founder of Khan Academy and Khanmigo and author and journalist Jacob Ward that can provide deeper engagement for teacher candidates in layering texts for critical inquiry (Table 2.1). Teacher educators may invite teacher candidates to consider dominant GenAI discourses by exploring how Sal Khan in a Ted Talk positioned GenAI as a personal tutor for students and personal assistant for teachers. Teacher candidates can examine how Khan characterizes the Socratic Seminar as a viable model for engaging with GenAI outputs, probe how this teaching model functions, and then critically reflect on Khan's teaching vision. Khan provides a demonstration of how Khanmigo operates as a Socratic platform that mimics conversational cognitive prompting, where students engage in dialogue with a literary persona, scientist, or prominent historical figure and collaborative writing to master skills.

Author/ Producer	Title	Link
Sal Khan	How AI ("Khanmigo") Could Save (Not Destroy) Education (TED Talk)	https://youtu.be/hJP5GqnTrNo?si=sPsIZ3ayYHYO2I3d
Jim Chilton	AI Will Reshape Education: Are We Building Tools We Can Trust? (TEDX)	https://www.youtube.com/watch?v=oV6HWzzeD-I
Natasha Berg	Should We Let Students Use ChatGPT?	https://www.youtube.com/watch?v=ogcSQ-cFRVM
UNESCO	Teacher to Teacher: Is AI Reshaping Education for the Better?	https://www.youtube.com/watch?v=EEbGYRW7feM
Sal Khan & Jacob Ward	Debate - School of Thought: Is AI Helping us Cheat or Learn?	https://www.youtube.com/watch?v=ApBNg1AzFE0

Table 2.1. GenAI resources for educators

8. Conclusion

Criticality is a foundational construct for teacher candidates as they develop balanced perspectives on language teaching while evaluating GenAI tools. This construct has sociopolitical implications, particularly for historically underrepresented learners. In this context, teacher candidates must gain a deep understanding of how GenAI tools can perpetuate challenges stemming from systemic injustices in language education. This process includes confronting negative impacts on Indigenous language programs, the affirmation of home and heritage languages, and the use of translanguaging as a humanizing practice, all while exacerbating the digital divide.

Ethical and responsible use of these tools requires an awareness of their capabilities and limitations in order to navigate GenAI platforms safely and cautiously. Without implementing protective measures both inside and outside the classroom, such as avoiding the use of personal and identifying information, fostering awareness of parasocial interactions, ensuring strong social-emotional support, minimizing overreliance on GenAI outputs for guidance and advice, and engaging with critical literacy dimensions (considering the sociocultural context, exploring multiple perspectives, and examining commonplace assumptions), teacher candidates and educators may experience harmful impacts on their professional autonomy and agency regarding their teaching practices and educational vision.

The phenomenon of anthropomorphizing GenAI chatbots—attributing human-like qualities to them (Williams et al., 2019)—is frequent and notable. This phenomenon is particularly applicable for digital agents used in student-facing GenAI tools and platforms. Pre-service and in-service teachers need to be aware that these chatbots can foster parasocial interactions during GenAI-mediated learning. Therefore, classroom activities should be carefully monitored for potential issues. Critical AI education should occur within a problem-posing context, where educators confront the implications of anthropomorphizing chatbots as a result of GenAI-mediated learning. Consequently, targeted social-emotional learning (SEL) lessons should be integrated as a preventive measure. This approach should include thoughtfully designed questions, activities, and reflections for students to engage with as they navigate these emerging contexts in which meaning making takes place.

For these reasons, teacher education and continuing education programs as well as school districts must provide timely preparation and guidance for teacher candidates before they take classroom positions in developing their pedagogical, content, and technical knowledge, underscored by criticality. To continue to develop their practice, teacher candidates could access professional learning and seminars through accessible platforms and networks such as their local Office of the Superintendent of Public Instruction, professional learning networks such as ACTFL, Coalition of Community Based-Heritage Language Schools, and university language centers, such as the Center for Educational Resources in Culture, Language and Literacy (CERCLL) at The University of Arizona to explore the most recent developments in GenAI applications in the language teaching field.

This chapter underlines the teacher-in-the-loop approach and envisions critical GenAI literacy practices that are culturally and linguistically responsive and aimed at justice-focused literacy for concurrently reading the word and world (Freire & Macedo, 2005). It also provides a thorough overview and analysis of relevant literature and approaches to GenAI literacy, specifically by applying Lewison et al.'s (2002) critical literacy framework to language teacher education. In the critical analysis of the scaffolded lesson plan and the suggested reflective practices, teacher agency and responsible GenAI integration are modeled and situated in critical AI literacy praxis and discussed through an ecocritical lens. This chapter is aimed at teacher candidates, in-service teachers, and teacher-educators, as well as district leaders and curriculum developers who consider it necessary to develop critical reflective practice alongside technological, pedagogical, and content knowledge. At the core of this practice is the need to understand not only the GenAI output but also what drives it and be able to critically analyze both the process and product concurrently through a critical lens.

Future research must consist of analyses of student-facing and teacher-facing platforms and the specific ways both groups are using and cognizing GenAI tools, platforms, and chatbots. These research endeavors should differentiate between using GenAI to develop content knowledge for identifying points of language growth, which constitutes process-oriented usage, and relying on GenAI for cognitive recall and retrieval, which constitutes product-oriented usage. More studies are needed that examine GenAI's role and effects on learners' procedural and conceptual knowledge in language development, particularly in concept practice and empowerment gained from cultivating knowledge and applying it to the real world. Qualitative longitudinal studies centering teachers and students are necessary to probe the long-term effects of using GenAI in language education with stronger emphasis on environmental justice and humanizing language practices that support equity, multilingualism, and sustainability.

References

Alasadi, E. A., & Baiz, C. R. (2023). Generative AI in education and research: Opportunities, concerns, and solutions. *Journal of Chemical Education, 100*(8), 2965-2971. https://doi.org/10/1021/acs.jchemed.3c00323

Belda-Medina, J., & Calvo-Ferrer, J. R. (2022). Using chatbots as AI conversational partners in language learning. *Applied Sciences, 12*(17), 8427. https://doi.org/10.3390/app12178427

Bender, E. M., Gebru, T., McMillan-Major, A., & Shmitchell, S. (2021, March). On the dangers of stochastic parrots: Can language models be too big? In *Proceedings of the 2021 ACM Conference on Fairness, Accountability, and Transparency* (pp. 610-623). https://doi.org/10.1145/3442188.3445922

Berg, N. (2023, September 24). *Should we let students use ChatGPT? TEDxSioux Falls* [Video]. YouTube. TEDx Talks. https://www.youtube.com/watch?v=ogcSQ-cFRVM

Bhaskar, P., & Seth, N. (2024). Environment and sustainability development: A ChatGPT perspective. In *Applied data science and smart systems* (pp. 54-62). CRC Press.

Blut, M., Wang, C., Wünderlich, N. V., & Brock, C. (2021). Understanding anthropomorphism in service provision: A meta-analysis of physical robots, chatbots, and other AI. *Journal of the Academy of Marketing Science, 49*, 632-658. https://doi.org/10.1007/s11747-020-00762-y

Chilton, J. (2025, April 10). *AI will reshape education. Are we building tools we can trust? TEDxSNHU* [Video]. YouTube. TEDx Talks. https://www.youtube.com/watch?v=oV6HWzzeD-I

Colleague AI. (2024, November 10). *Privacy policy.* https://www.colleague.ai/privacy-policy/

Cope, B., Kalantzis, M., & Searsmith, D. (2021). Artificial intelligence for education: Knowledge and its assessment in AI-enabled learning ecologies. *Educational philosophy and theory, 53*(12), 1229-1245. https://doi.org/10.1080/00131857.2020.1728732

Corp, A., & Revelle, C. (2023). ChatGPT is here to stay: Using ChatGPT with student teachers for lesson planning. *The Texas Forum of Teacher Education, 14,* 116-124. https://txate.org/resources/Documents/Forum%202023%20Final%20Full%20Issue.pdf#page=122

De la Vall, R. R. F., & Araya, F. G. (2023). Exploring the benefits and challenges of AI-language learning tools. *International Journal of Social Sciences and Humanities Invention, 10*(01), 7569-7576. https://doi.org/10.18535/ijsshi/v10i01.02

de Roock, R. S. (2024). To become an object among objects: Generative artificial "intelligence," writing, and linguistic white supremacy. *Reading Research Quarterly, 59*(4), 590-608. https://doi.org/10.1002/rrq.569

Dilek, M., Baran, E., & Aleman, E. (2025). AI literacy in teacher education: Empowering educators through critical co-discovery. *Journal of Teacher Education, 76*(3), 294-311. https://doi.org/10.1177/00224871251325083

Diliberti, M., Schwartz, H. L., Doan, S., Shapiro, A. K., Rainey, L., & Lake, R. J. (2024). *Using artificial intelligence tools in K-12 Classrooms.* Rand.

Doha Debates. (2023, November 14). *School of thought: Is AI helping us cheat or learn?* [Video podcast episode]. YouTube. https://www.youtube.com/watch?v=ApBNg1AzFE0

Doo, F. X., Vosshenrich, J., Cook, T. S., Moy, L., Almeida, E. P., Woolen, S. A., Gichoya, J. W., Heye, T., & Hanneman, K. (2024). Environmental sustainability and AI in radiology: a double-edged sword. *Radiology, 310*(2), e232030. https://doi. org/10.1148/radiol.232030

Dwyer-Kuntz, T., & Kuntz, S. (2024). Spells for success: Promoting UDL and DI with MagicTools. *Artificial intelligence in education conference: Shaping future classrooms.* eCampusOntario. https://ecampusontario.pressbooks.pub/artificialintelligenceineducationconference/chapter/spells-for-success-promoting-udl-and-di-with-magictools/

Freire, P., & Macedo, D. (2005). *Literacy: Reading the word and the world.* Routledge. https://doi.org/10.4324/9780203986103

García, O. (2013). From diglossia to transglossia: Bilingual and multilingual classrooms in the 21st century. *Bilingual and multilingual education in the 21st century: Building on experience, 94.* https://doi.org/10.2307/jj.27080031

García, O., & Kleifgen, J. A. (2020). Translanguaging and literacies. *Reading Research Quarterly, 55*(4), 553–571. https://doi.org/10.1002/rrq.286

George, A. S., George, A. H., & Martin, A. G. (2023). The environmental impact of AI: a case study of water consumption by ChatGPT. *Partners Universal International Innovation Journal, 1*(2), 97-104.https://doi.org/10.5281/zenodo.7855594

Giroux, H. (1985). Teachers as transformative intellectuals. *Social Education, 49*(5), 376–379. https://www.taylorfrancis.com/chapters/edit/10.4324/9780429495670-16/teachers-transformative-intellectuals-henry-giroux

Guerrettaz, A. M., Zahler, T., Sotirovska, V., & Boyd, A. S. (2022). 'We acted like ELLs:' A pedagogy of embodiment in preservice teacher education. *Language Teaching Research, 26*(6), 1274-1298. https://doi.org/10.1177/1362168820909980

Hargrave, M., Fisher, D., & Frey, N. (2024). *The Artificial Intelligence Playbook: Time-Saving Tools for Teachers that Make Learning More Engaging.* Corwin Press. https://www.google.com/books/edition/The_Artificial_Intelligence_Playbook/YvPzEAAAQBAJ?hl=en

Henriksen, D., Creely, E., Gruber, N., & Leahy, S. (2025). Social-emotional learning and generative AI: A critical literature review and framework for teacher education. *Journal of Teacher Education, 76*(3), 312-328. https://doi.org/10.1177/00224871251325058

Hurley, Z. (2024). Generative AI's family portraits of whiteness: A postdigital semiotic case study. *Postdigital Science and Education, 6*(4), 1240-1260. https://doi.org/10.1007/s42438-024-00491-3

Hwang, G. J., Xie, H., Wah, B. W., & Gašević, D. (2020). Vision, challenges, roles and research issues of artificial intelligence in education. *Computers and Education: Artificial Intelligence, 1,* 100001. https://doi.org/10.1016/j.caeai.2020.100001

Kalantzis, M., & Cope, W. (2023). Multiliteracies: Life of an idea. *The International Journal of Literacies, 30*(2), 17. https://doi.org/10.18848/2327-0136/CGP/v30i02/17-89

Khan, S. (2023, May 1). *How AI could save (not destroy) education. TED*[Video]. YouTube. TEDx Talks. https://www.youtube.com/watch?v=hJP5GqnTrNo

Kilbane, C. R., & Milman, N. B. (2013). *Teaching models: Designing instruction for 21st century learners.* Pearson Higher Ed.

Kim, M., & Adlof, L. (2024). Adapting to the future: ChatGPT as a means for supporting constructivist learning environments. *TechTrends, 68*(1), 37-46. https://doi.org/10.1007/s11528-023-00899-x

Kohnke, L., & Zou, D. (2025). Artificial intelligence integration in TESOL Teacher education: Promoting a critical lens guided by TPACK and SAMR. *TESOL Quarterly*. https://doi.org/10.1002/tesq.3396

Kohnke, L., Moorhouse, B. L., & Zou, D. (2023). ChatGPT for language teaching and learning. *Relc Journal, 54*(2), 537-550. https://doi.org/10.1177/0033 6882231162868

Kostka, I. (2023). Using a flipped learning approach to teach source-based writing in academic English language courses. In *Teaching and learning source-based writing* (pp. 94-107). Routledge. https://www.taylorfrancis.com/chapters/edit/10.4324/9781003283485-9/using-flipped-learning-approach-teach-source-based-writing-academic-english-language-courses-ilka-kostka

Kostka, I., & Marshall, H. W. (2018). Flipped learning in TESOL: Past, present, and future. In *Applications of CALL theory in ESL and EFL environments* (pp. 223-243). IGI Global Scientific Publishing. https://doi.org/10.4018/978-1-5225-2933-0.ch013

Kostka, I., & Toncelli, R. (2023). Exploring applications of ChatGPT to English language teaching: Opportunities, challenges, and recommendations. *Tesl-Ej, 27*(3), n3. https://doi.org/10.55593/ej.27107int

Kousa, P., & Niemi, H. (2023). AI ethics and learning: EdTech companies' challenges and solutions. *Interactive Learning Environments, 31*(10), 6735-6746. https://doi.org/10.1080/10494820.2022.2043908

Leander, K. M., & Burriss, S. K. (2020). Critical literacy for a posthuman world: When people read, and become, with machines. *British Journal of Educational Technology, 51*(4), 1262-1276. https://doi.org/10.1111/bjet.12924

Lewison, M., Flint, A. S., & Van Sluys, K. (2002). Taking on critical literacy: The journey of newcomers and novices. *Language arts, 79*(5), 382-392. https://doi.org/10.58680/la2002255

Liu, A., & Sun, M. (2023). From voices to validity: Leveraging large language models (LLMs) for textual analysis of policy stakeholder interviews. *AERA Open, 11*. https://doi.org/10.1177/23328584251374595

Lo, C. K. (2023). What is the impact of ChatGPT on education? A rapid review of the literature. *Education Sciences, 13*(4), 410. https://doi.org/10.3390/educsci13040410

Long, D., & Magerko, B. (2020, April). What is AI literacy? Competencies and design considerations. In *Proceedings of the 2020 CHI conference on human factors in computing systems* (pp. 1-16). https://doi.org/10.1145/33138 31.3376727

Luckin, R., Holmes, W., Griffiths, M., & Forcier, L. B. (2016). *Intelligence Unleashed: An Argument for Ai in Education*. Pearson London. https://discovery.ucl.ac.uk/id/eprint/1475756/

Luke, A. (2012). Critical literacy: Foundational notes. *Theory into Practice, 51*(1), 4-11. https://doi.org/10.1080/00405841.2012.636324

Ma, Q., Crosthwaite, P., Sun, D., & Zou, D. (2024). Exploring ChatGPT literacy in language education: A global perspective and comprehensive approach. *Computers and Education: Artificial Intelligence, 7,* 100278. https://doi.org/10.1016/j.caeai.2024.100278

MagicSchool. (n.d.). *FAQs.* Retrieved October 20, 2025 from https://www.magicschool.ai/faq

Miao, F., & Cukurova, M. (2024). *AI competency framework for teachers.* UNESCO. Retrieved July 24, 2025 from https://doi.org/10.54675/ZJTE2084

Mishra, P., & Koehler, M. J. (2006). Technological pedagogical content knowledge: A framework for teacher knowledge. *Teachers College Record, 108*(6), 1017-1054. https://doi.org/10.1111/j.1467-9620.2006.00684.x

Moorhouse, B. L. (2024). Beginning and first-year language teachers' readiness for the generative AI age. *Computers and Education: Artificial Intelligence, 6,* 100201. https://doi.org/10.1016/j.caeai.2024.100201

Moorhouse, B. L., & Kohnke, L. (2024). The effects of generative AI on initial language teacher education: The perceptions of teacher educators. *System, 122,* 103290. https://doi.org/10.1016/j.system.2024.103290

Muhammad, G. (2020). *Cultivating genius.* Scholastic.

Munro, R. (2021). *Human-in-the-loop machine learning: Active learning and annotation for human-centered AI.* Manning.

Ng, D. T. K., Leung, J. K. L., Chu, S. K. W., & Qiao, M. S. (2021). Conceptualizing AI literacy: An exploratory review. *Computers and Education: Artificial Intelligence, 2,* 100041. https://doi.org/10.1016/j.caeai.2021.100041

Otheguy, R., & García, O. (2024). Translanguaging. In *Oxford Research Encyclopedia of Linguistics.* https://doi.org/10.1093/acrefore/9780199384655.013.1031

Paiz, J. M., Toncelli, R., & Kostka, I. (2025). *Artificial intelligence, real teaching: A guide to AI in ELT.* University of Michigan Press. https://doi.org/10.3998/mpub.14329762

Park, Y. S. (2024). White default: Examining racialized biases behind AI-generated images. *Art Education, 77*(4), 36-45. https://doi.org/10.1080/00043125.2024.2330340

Peñalvo, F. J. G., & Ingelmo, A. V. (2023). What do we mean by GenAI? A systematic mapping of the evolution, trends, and techniques involved in Generative AI. *IJIMAI, 8*(4), 7-16. https://doi.org/10.9781/ijimai.2023.07.006

Sardinha, T. B. (2024). AI-generated vs human-authored texts: A multidimensional comparison. *Applied Corpus Linguistics, 4*(1), 100083. https://doi.org/10.1016/j.acorp.2023.100083

Shieh, E., Vassel, F. M., Sugimoto, C., & Monroe-White, T. (2024). Laissez-faire harms: Algorithmic biases in generative language models. *arXiv preprint arXiv:2404.07475.* https://doi.org/10.1609/aies.v8i3.36722

Sotirovska, V. (2024). Critical visual analysis of three multimodal texts about displacement for classroom practice. *Dragon Lode, 42*(1), 29-42. https://www.clrsig.org/uploads/1/1/5/0/115059369/critical_visual_analysis_of_three_multimodal_texts_about_displacement_for_classroom_practice_-_sotrovska.pdf

Sotirovska, V. (2025). An arts-based multiliteracies approach to exploring international art-themed picture books. In *Arts-based multiliteracies for teaching and learning* (pp. 409-442). IGI Global. https://doi.org/10.4018/979-8-3693-3184-2.ch014

Sotirovska, V., & Kelley, J. (2020). Anthropomorphic characters in children's literature: Windows, mirrors, or sliding glass doors to embodied immigrant experiences. *The Elementary School Journal, 121*(2), 337-355. https://doi.org/10.1086/711054

Sotirovska, V., & Vaughn, M. (2022). Developing preservice teachers' critical literacy praxis in a rural teacher education program. *Reading Horizons: A Journal of Literacy and Language Arts, 61*(2), 2. https://scholarworks.wmich.edu/reading_horizons/vol61/iss2/2

Sotirovska, V., & Vaughn, M. (2023a). Examining pre-service teachers' critical beliefs: Validation of the critical literacy beliefs survey (CLBS). *Teaching Education, 34*(2), 170-193. https://doi.org/10.1080/10476210.2022.2049742

Sotirovska, V., & Vaughn, M. (2023b). *Enhancing educators' theoretical and practical understandings of critical literacy.* Cambridge University Press. https://doi.org/10.1017/9781009304726

Souto-Manning, M., & Martell, J. (2016). *Reading, writing, and talk: Inclusive teaching strategies for diverse learners, K–2.* Teachers College Press.

Sperling, K., Stenberg, C. J., McGrath, C., Åkerfeldt, A., Heintz, F., & Stenliden, L. (2024). In search of artificial intelligence (AI) literacy in Teacher education: A scoping review. *Computers and Education Open,* 100169. https://doi.org/10.1016/j.caeo.2024.100169

Stolpe, K., & Hallström, J. (2024). Artificial intelligence literacy for technology education. *Computers and Education Open, 6,* 100159. https://doi.org/10.1016/j.caeo.2024.100159

Toncelli, R., & Kostka, I. (2024). A love-hate relationship: Exploring faculty attitudes towards GenAI and its integration into teaching. *International Journal of TESOL Studies, 6*(3), 77-94. https://doi.org/10.58304/ijts.20240306

Tram, N. H. M. (2025). Unveiling the drivers of AI integration among language teachers: Integrating UTAUT and AI-TPACK. *Computers in the Schools, 42*(2), 100-120. https://doi.org/10.1080/07380569.2024.2441155

UNESCO. (2025, January 23). *Teacher to teacher: Is AI reshaping education for the better?* [Video]. YouTube. https://www.youtube.com/watch?v=EEbGYRW7feM

University of Washington College of Education Newsroom. (2024, October 9). *Interdisciplinary, cross-sector team led by UW's Dr. Min Sun awarded $10 million IES grant to launch AmplifyGAIN Center.* Retrieved July 24, 2025 from https://education.uw.edu/news/feature/interdisciplinary-cross-sector-team-led-uws-dr-min-sun-awarded-10-million-ies-grant

Uwosomah, E. E., & Dooly, M. (2025). It is not the huge enemy: preservice teachers' evolving perspectives on AI. *Education Sciences, 15*(2), 152. https://doi.org/10.3390/educsci15020152

WIDA [World-Class Instructional Design and Assessment]. WIDA English language development standards framework, 2020 edition: Kindergarten-Grade 12. https://wida.wisc.edu/sites/default/files/resource/WIDA-ELD-Standards-Framework-2020.pdf

Williamson, B. (2024). The social life of AI in education. *International Journal of Artificial Intelligence in Education, 34*(1), 97-104. https://doi.org/10.1007/s40593-023-00342-5

Zhu, M., & Wang, C. (2025). A systematic review of research on AI in language education: Current status and future implications. *Language Learning & Technology, 29*(1), 1-29. https://doi.org/10.2139/ssrn.4684304

Appendix A:
lesson plan annotations for teacher preparation classroom modeling

In each section of this ChatGPT-generated lesson plan. I have included Pause and Reflect sections (1-8) with guiding questions and prompts, inviting teacher candidates to consider how they would contextualize, modify, substitute, adapt, or transform elements of this lesson plan. Teacher candidates are also encouraged to reflect on the example adaptations included in the lesson scaffold. To prompt reflection, teacher candidates could begin engaging with the suggested prompts as a point of departure in their discussion.

ChatGPT-Generated Response with Teacher Input in Italicized Text

Lesson Plan: constructing Language Arts Narratives in a Multilingual Classroom (ChatGPT-generated Response with Teacher Input in Italicized Text)

Subject: English Language Development (ELD)

Grades: 2-3

Duration: 60 minutes

WIDA Standard: ELD-LA.2-3. Narrate. Expressive: multilingual learners construct language arts narratives that

- Orient audience to context
- Develop story with time and event sequences, complication, resolution, or ending
- Engage and adjust for audience

Objectives:

1. Students will be able to construct a narrative that orients the audience to the context.

Objective 1 addresses the teaching context and will be the point of departure for the analysis of this lesson plan.

Pause & Reflect 1
Think about your teaching context, who your students are, and what is relevant to their lives. Reflect on the sociocultural, demographic, and intersectional contexts of your classroom. This is key information that only teachers know through conferring, listening sessions, and daily interactions with students, families, and communities and such engagements result in deep knowledge of students' unique socioemotional characteristics, needs, interests, prior knowledge, funds of knowledge, and full unitary repertoire (García, 2013).
Ask yourself: who is this lesson designed for?
• *Read over the lesson plan objectives and notice the language used in each objective. Does it meet students where they are and does it apply to who they are as human beings and then learners? Please explain.*
• *The first half of Objective 2 details a story-based task focused on understanding a sequence of events, chronology in storytelling, narrative sentence structures, and verb tense frames. The second half of O2 also details the conflict and the resolution within this story-based writing task.*

2. Students will develop a story with a clear sequence of events, including a complication and a resolution.

Pause & Reflect 2
Evaluate the lesson for language demands, including the language structures and genre. Consider presenting students with a mentor text in the form of a mini lesson so that they can visualize the task (e.g., show students an annotated example and perform a think aloud to showcase the thought process steps required for this task). After evaluating the language demands, consider providing additional scaffolds for this mini lesson, such as environmental print, anchor chart, or semantic map, scaffolding layer of literacy support. For example, consider including dual language sentence stems as language supports and an anchor chart to show the sequence of events as an example of a narrative arc along with mentor text responsive to the classroom context. Consider how adding a dual language picture book can support the mini lesson you are teaching. The more diverse the resources are, the better.

How can you plan for deeper learning?

- *Objective 3 focuses on the expressive aspect of the task, specifically the presentation mode. Teachers may choose to add multiple means of expression in addition to the traditional ones (reading and writing) as presentation modes. Consider what other modes you can include to layer multimodal supports (Muhammad, 2020) to create depth of learning.*

3. Students will learn to engage their audience and adjust their narrative for different audiences.

Materials

- Storybooks reflecting diverse cultures and languages (e.g., "The Name Jar" by Yangsook Choi, "My Name is Yoon" by Helen Recorvits)
- Graphic organizers (story maps)
- Visual aids (pictures, illustrations)
- Bilingual dictionaries
- Writing journals
- Digital storytelling tools (e.g., Storybird, Book Creator)
- Examples of culturally diverse narratives

Pause & Reflect 3

How can you plan for deeper learning?

- *Consider introducing dual language picture books for a multiliteracies, multilingual, and multimodal approach (Sotirovska & Vaughn, 2023b). This approach engages students in layering texts in various modalities, engage in presenting through storytelling, and present their literacy artifacts within the class community. Facilitate multilingual storytelling in multiple students' languages, such as their home and heritage languages and the language of instruction.*

Resources:

- Access to a classroom library with culturally diverse books
- Online resources for digital storytelling
- Bilingual support staff or volunteers (if available)

- Visual aids and story maps templates

Pause & Reflect 4
What learning scaffolds can you create to support students?
• *A video-recorded mini lesson is a scaffold for students, which provides multiple points of access to the content, instructions, and examples, enabling students to replay the video for comprehension of the learning targets and lesson instructions.*

Lesson Plan:

Introduction (10 minutes):

1. **Greeting and Warm-Up:**
    - Greet students in various languages represented in the classroom.
    - Quick icebreaker: "Tell us about a favorite story from your culture."
2. **Objective Overview:**
    - Explain the objectives of the lesson.
    - Briefly discuss what makes a good story: context, sequence of events, complication, resolution, and audience engagement.

Pause & Reflect 5
What opportunities do you see for culturally and linguistically responsive meaning-making?
• *Consider the following activity: students work in groups or in pairs to answer: What makes a good story? Who are the storytellers in our community? What is storytelling? Who in the community can we learn from?*

Centering Indigenous Community Literacies and Place-based Education Models

- *For example, in Washington state, you can layer lessons centering on Indigenous Ways of Knowing and explore John McCoy (lulilas) the state Indigenous curriculum Since Time Immemorial curriculum by including oral storytelling artifacts (e.g., Father Ocean and How Fire Came to Earth). Then ask yourself, what kind of literacy artifacts can you include in your teaching? Who are the cultural ambassadors, stewards, and community voices in your context of teaching and learning context?*
- *This is a culturally responsive and sustaining way to engage students in the cultural funds, histories, and wealth of communities by enacting place-based education models. Turn to your community and think about who you can learn from.*

Instruction (20 minutes):

1. **Read Aloud:**
    - Read a storybook that reflects a culture represented in the classroom. For example, "The Name Jar" by Yangsook Choi.
    - Discuss the elements of the story: context, events, complication, resolution.
2. **Story Mapping:**
    - Introduce the story map graphic organizer.
    - Model how to fill in the story map using the read aloud story.
3. **Group Discussion:**
    - Discuss how the story engages the audience and how the author adjusts the narrative for different readers.

Guided Practice (15 minutes):

1. **Story Map Creation:**
    - In pairs or small groups, students choose a cultural story they know and fill out a story map.
    - Encourage use of bilingual dictionaries and visual aids for support.

Pause & Reflect 6
What metacognitive strategies can support learning?
• *Consider encouraging students to create a visual representation of their inquiry timeline, entitled a Story of Wonderings. Ask them to include the following: What was the writing process like for them? Who did they talk to? What object/artifact (e.g., a photograph/video/picture book/an object) inspired their story? Support this skill across students' full unitary repertories.* • *Encourage students to use author journals instead of worksheets and pre-made graphic organizers. An author's journal is a notebook designated for writing. Students can be encouraged to create their own graphic organizers in their journals and use annotations to add gradual information as they are making sense of the storytelling practice.*

2. **Cultural Sharing:**
   - Each group shares their story map with the class.
   - Discuss similarities and differences in narrative structure across cultures.

Independent Practice (10 minutes):

1. **Writing Own Narrative:**
   - Students begin drafting their own narrative in their writing journals using their story map.
   - Encourage students to include cultural elements and use their home language if needed.

Closing (5 minutes):

1. **Sharing and Feedback:**
   - A few students volunteer to share their drafts.
   - Provide positive feedback focusing on context, sequence, and engagement.
2. **Homework Assignment:**
   - Finish writing the narrative at home.
   - Practice telling the story to a family member or friend and note any feedback.

Pause & Reflect 7

Which asset-based assessment practices can support students' learning experiences?

- *Encourage peer reviews and facilitate individual writing conferences with the students. Ask students to share their stories with family, community members, or friends as thought partners. This practice will help cultivate connections and student communities.*

- *Everyone should have the opportunity to share and engage with their classmates' content. Consider facilitating a Think-Pair-Share or a group/circle share out. As a reflection activity, students might share favorite parts of their stories. They can also discuss favorite moments or highlight their most memorable experience in the storytelling inquiry process.*

Assessment:

- **Formative Assessment:**
  - Observe students during group discussions and story mapping.
  - Provide feedback during guided and independent practice.
- **Summative Assessment:**
  - Evaluate the final narratives based on a rubric that includes context orientation, sequence of events, complication and resolution, and audience engagement.

Reflection:

- After the lesson, reflect on the effectiveness of the activities and materials.
- Consider how well students were able to connect their cultural backgrounds to the narrative structure.
- Plan for adjustments in future lessons to better support diverse learners and enhance culturally responsive pedagogy.

Extension Activities:

- **Digital Storytelling:**

  - Students can use digital tools like Storybird or Book Creator to publish their narratives.
  - Share digital stories with the class or a wider audience.
- **Cultural Storytelling Event:**
  - Organize a storytelling event where students can share their narratives with parents and community members.
  - Invite guest storytellers from different cultures to share their stories.

This lesson plan leverages culturally responsive and sustaining pedagogies to create an inclusive and engaging learning environment for multilingual students. It emphasizes the importance of cultural context in narrative construction and provides multiple entry points for students to connect their personal experiences with the learning objectives.

Pause & Reflect 8
Which practices can support making learning visible, joyful, and communal?
• *Consider organizing student-led portfolio conferences, where students take ownership of the writing process by documenting how they developed their story through drafting, interviewing, annotating, and gathering information.* • *Consider encouraging students to reflect on what they learned about their families and/or communities, discovered about themselves, and help them make connections between these experiences and the five HRL pursuits, Identity, Skills, Intellect, Criticality, and Joy (Muhammad, 2020).* • *Sharing students' stories with the school community is a great way to situate their learning in the real world in an event entitled, Author Voices Heard & Seen. Consider inviting other students from different grade levels, as well as parents, guardians, and caregivers, to attend this event.*

Appendix B: probing questions for developing AI literacy

Question Categories	Sample Questions
Disrupting the Commonplace	What do I know about GenAI? (e.g., Teacher candidates can first probe their prior knowledge of GenAI and describe how they perceive this technology.) What do I know about GenAI in language teaching? What do I want to know about GenAI in language teaching? What content, pedagogical, and technical knowledge do I need to critically engage with GenAI outputs? Where have I observed GenAI being used in language education the most and for what? (e.g., Candidates could create visual maps of GenAI usage, discuss cases of use, describe school contexts and/or their own for GenAI use). In what ways do I envision GenAI being helpful and/or challenging in my teaching preparation? What are some examples? Now consider the following: how do I begin to form a stance (e.g., pedagogical beliefs and opinions) about GenAI?
Focusing on Sociopolitical Issues	What are the self-perceived advantages and disadvantages of using GenAI in your community of teaching and learning? Please provide examples from your own experiences and observations. What are some significant examples of how GenAI has changed language education since its introduction (e.g., ChatGPT in 2022)? In what way, if any, are you engaging with GenAI? How are you learning about this technology? In what way has GenAI been introduced and/or applied in your school contexts, whether that be in your teacher education classes, practicum experiences, and/or professional community? Now consider the following: how do I begin to engage in problem-posing inquiry about GenAI?

Exploring Multiple Perspectives	What are some differing opinions on GenAI in the language education field? Whose ethical, pedagogical, and technological perspectives on GenAI shape your beliefs about how this technology is applied to language education? To better understand the impact on schooling, what are community members and stakeholders saying about GenAI (e.g., other teachers, students, and families, etc.)? What are teachers saying? What are students saying? What are parents, caregivers, and/or guardians saying? Now consider the following: how do I begin to engage a critical inquiry approach in my teaching and learning practice?
Taking Action to Promote Social Justice	As GenAI is being integrated into education systems, how is this change reshaping existing power structures? How is it affecting the power dynamics between teachers, students, and administrators? How can GenAI influence teacher agency and give power to technology and technology providers? How would this affect student learning? What are some of the ways GenAI usage can promote/reduce inequities in the language education classroom? What are some actions you can take to ensure you are teaching with agency (e.g., making curricular choices and decisions and influencing outcomes in how and the extent to which GenAI is used in teaching and learning)? Now consider the following: how do I begin to transform/influence/adapt GenAI tools, outputs, and approaches for educational equity in my own community of practice?
Critical Noticing	What themes and patterns can you observe in GenAI outputs? In what ways can GenAI outputs be adapted for your unique learning and teaching context? (e.g., Candidates can explore data sets and engage in a Socratic seminar lesson model to structure a discussion around their inferences after exploring GenAI outputs.)

Critical Visioning	What role does GenAI currently play in language teaching and learning? What are some examples? How do you envision your role as a teacher changing in the GenAI age? What are some examples of a task, responsibility, or way you approach teaching and learning that has or might change? How are students' roles in their learning evolving in the GenAI age? Could you describe a classroom situation where you have noticed this change? Considering your responses to the previous question, what is your vision for GenAI in language education?
Layering Texts Practice	How can you ensure that you are layering multiple perspectives alongside GenAI outputs, such as including primary and secondary sources? What instructional steps are you taking to ensure credible sources are introduced?
Critical Communities of Practice	Is GenAI use reflective of multiliteracies (multiple types of literacy), multimodality (multiple forms of communication, such as text, image, video), and multilingualism (multiple languages, such as including students' home and heritage languages or dual language artifacts, such as picture books being recommended for teaching)? What are some examples? Are all stakeholders (e.g., teachers, students, administrators, parents, guardians) modeling responsible and situated use of GenAI tools? How does this practice help shape a critical community where you teach and learn? What are some examples of said practices? Is critical reflection on the effects of GenAI on language teaching and learning discussed at your school of work/practicum/teacher education? What does this practice look like in your critical community?

Inquiry into Instructional Planning	How do GenAI outputs contribute to new understandings and enduring pursuits? How do GenAI outputs contribute to new understandings of content knowledge in language teaching?
Curriculum Alignment	How do GenAI outputs account for grade level clusters in language teaching? In what way do GenAI outputs reflect curriculum scope and sequence in language planning? To what extent do GenAI outputs reflect unpacked and contextualized language standards?
Critical Information Literacy	What instructional strategies can help ensure that GenAI outputs do not contribute and/or perpetuate false, biased, or incomplete information?
Visioning and Reflective Practice	How are you developing and maintaining critical praxis (connecting theory with practice through) in a time of heighted GenAI popularity? How do these practices help you be more critical and reflective about your teaching and learning (e.g., lesson planning, text selection, student activities, instructional strategies)? What are some examples?
Inquiry into Instructional Planning	How does integrating GenAI content into lessons help students develop a comprehensive understanding of the learning objectives, if at all?
High-Leverage Practices	Does the generated output provide guidance for differentiation? If so, how? Does the GenAI platform support multiple ways of content access, representation, and expression?) Does the generated output provide culturally and historically situated examples, assessments, teaching materials, and tools? If so, in what ways and what are some examples?

Incorporating Scholarly Practices	Are the resources cited in GenAI outputs peer reviewed, appear in notable publications, and/or verified through existing works? (Ensure that you are tracing claims to original peer-reviewed sources, corroborating information with original peer reviewed sources, and verifying the content for accuracy, representation, and contextualization) How can you ensure that the generated content is free of GenAI hallucinations? How will you know?
Critical Pedagogy Context	How do you ensure your lessons create connections to the real world? What are some examples of practice? How do you ensure that your lessons reflect the current socio-political-cultural contexts relevant to students' lives? What are some examples of practice? What practical adaptations would you need to make to invite Criticality, in the form of sociopolitical awareness, (Muhammad, 2020) in your teaching and learning?

Chapter 3

Digital storytelling across cultures: empowering multilingual minds

Ömer Buğra Hamzaoğlu
Mediterranean (Akdeniz) University

Başak Eda Hancı-Azizoğlu
Mediterranean (Akdeniz) University

Abstract

In second language learning, there is a need for a more interactive and culturally sensitive approach to teaching for multilingual learners. One way to achieve this is digital storytelling, a method that can help teachers establish interactive and empowering learning environments. This chapter reviews the existing literature by exploring how digital storytelling can be used as a pedagogical tool that enhances language learning through culturally inclusive narratives. Digital tools, such as artificial intelligence and virtual reality, are also discussed, emphasizing their potential to personalize and improve the storytelling experience for language learners. This chapter explores the efficacy of digital storytelling and promotes linguistic and cultural awareness in diverse classroom settings.

Keywords: Second language acquisition, Cultural narrative, Artificial intelligence in education, Virtual reality in education, Language learning pedagogy

* * *

1. Introduction

Digital storytelling is a powerful learning activity that has the potential to enhance language acquisition through meaningful stories. It is an important teaching approach in the world of language where the need for culturally methodological approaches is in high demand as it becomes increasingly and more diverse and intertwined. In simple words, digital storytelling is the act of using digital media creatively to convey tailored narratives. In various educational settings such as higher education, professional training, and adult learning, it has been shown to have significant instructional value for teachers and learners (McGee, 2014). By allowing learners to share their unique experiences and perspectives through the medium of digital storytelling, learning environments can be used to establish a more personal, deeper engagement with the language itself.

Methods like digital storytelling are also growing in popularity due to the increasing interest in the advancements that are happening in the world of technology. Digitals tools and platforms such as artificial intelligence (AI) and virtual reality (VR) systems is expanding in the context of language education. On the side of teachers, a wide range of language methods and strategies are being used with these digital tools to enhance the quality of teaching in terms of creating a reflective interaction, alongside meaningful conversations and narratives for their students. This innovative approach enables learners to explore their language skills in a dynamic and interactive environment. It acts as a pedagogical tool that captures learners' attention and fosters meaningful context for language acquisition. Teachers are able to create such engaging learning environments that resonate with the preferences of digitally native students by integrating visual and auditory elements into the storytelling process (Robin, 2008). This approach allows language learners to actively engage with the content. Digital storytelling not only boosts learners' language skills, but also their creativity and critical thinking abilities (Yang et al., 2022). In other words, digital storytelling is a powerful tool that enhances student motivation, improves comprehension and promotes collaborative learning, making it a valuable asset in the pursuit of effective language teaching and learning.

Language learners often engage with digital media and platforms throughout their daily lives. They are the generation of digital learners, growing up in a world where physical and virtual spaces coexist. Many language learners come to school being multilingual and having engaged with different cultures all around the world due to the prevalence of the digital world. In the context of linguistic and sociolinguistic themes, this engagement is also a part of their learning process, in which learners use and interact with the authentic language outside of their classroom. For teachers, it is essential to incorporate these interactions into the classroom settings and other learning environments. Since learners have limited knowledge of the target language, incorporating a digital approach to pedagogical reflection can be especially beneficial in second language learning settings. By utilizing digital media in language learning environments, learners have opportunities to interact with various forms of linguistic and sociolinguistic content, leading to a deeper understanding of the language and culture. Additionally, the implementation of this digital pedagogical reflection fosters collaboration amongst learners, providing linguistic insight, and helping them develop their linguistic negotiation skills (Guillén et al., 2020). This approach also enables language learners to practice and apply their skills in a supportive environment and adapt to language changes that occur in digital platforms, as well as developing and expanding the language proficiency of second language learners (Androutsopoulos, 2021).

This chapter explores the existing literature on multilingual learners and their cultural-linguistic identities in the context of second language learning. This study proposes practical teaching strategies for language learners using digital storytelling. It reviews the literature on the efficacy of digital storytelling for language learning, highlighting its potential to enhance learner engagement, motivation, and collaboration for the sake of cultural understanding amongst language learners. By reviewing insights from various studies, this chapter presents a framework for the implementation of digital storytelling in language classrooms, providing considerations for teachers working with multilingual learners.

2. Multilingual learners of the digital age

The latest generation in this digital world is much more comfortable in digital spaces compared to their parents, who may be the same age as their teachers. They want to maintain their digital lifestyles even within the school environment, incorporating their learning into it. The children of this digital age are also autonomous learners, and they take pride in the fact that they have the ability to be the architects of their learning journeys. This desire to control the learning process reflects the same mentality as in video games – the players are the heroes who define not just their actions in the game, but also the logic behind the game itself. The digital world is governed by similar rules. Once immersed in the digital world, these children expect a certain degree of freedom—a notion that is both valued and criticized. Consequently, their ideal teacher is not an authority figure, but rather someone who inspires and facilitates learning (Anderson & Macleroy, 2020). These digital learners are fundamentally different from the previous generation of students. They are not just surrounded by technology; they were born into it. Interaction with digital devices and media occurs on a regular basis for them. For this reason, a teaching approach such as digital storytelling is an attractive medium for young learners. According to the 2023 Ofcom report, children begin using the Internet at the age of three, and the gap between the age at which a child begins to use the Internet and watch television has decreased over the years. According to the same report, before starting school, children all around the world have access to computers. In other words, integration of digital tools into educational settings is not a necessity, but rather an inevitability.

2.1. Digital proficiency of young learners

Children's digital proficiency may vary from one learner to another since the concept of digital proficiency generally refers to the skills of using digital media and platforms, and it is affected by various conditions such as age, gender, parents' occupation and their educational level, etc. Although the reasons behind it may be socioeconomic and cultural, young multilingual children are known to have a high level of curiosity, along with self-confidence. The attributes can help them improve their aptitude towards any form of digital tool (Liang et al., 2021).

Understanding the child's digital proficiency helps language teachers analyze both their own digital competencies and that of their students with ease Likewise, adult language learners can be at different levels regarding their digital proficiencies (Kincl & Štrach, 2021). Therefore, prior experiences with digital storytelling can vary among students; therefore, two distinct criteria for language learners from a language teacher's perspective come into play. The first criterion is the students not having any sort of previous interaction with the digital tool might need more time in the introduction phase. It is crucial for them to familiarize themselves with the tool itself, alongside other factors such as time limitations, before proceeding to the topic they are going to learn. Secondly, although children with prior experience may require less time to familiarize themselves with technological equipment, they may still refrain from participating unless the teacher selects topics and tools that meaningfully engage their interests. Therefore, it is essential to establish a connection between students' preferences and interests, in order to keep them engaged and motivated throughout the process (Rahiem, 2021). It is pivotal for language teachers to understand a child's digital proficiency, not just for their growth and development in the digital landscape, but also to guide their interactions with digital tools effectively. By recognizing these varying levels, teachers can ensure that all students have the opportunity to engage in storytelling.

2.2. Impact of digital devices in classrooms

Creating a happy and comfortable classroom environment would help teachers gain the students' trust more easily and help them motivate their students to embrace the target language learning. Consequently, a balanced and reasonable use of digital devices combined with clear learning purposes should encourage teachers to seize opportunities for sustained language development (Kristiawan et al., 2022). Even though most children and young adults are digital natives, each learner requires guidance from their teacher to help them find a balance in their use of such devices, including controlling use during class hours. According to Kendrick et al. (2022), teachers using digital devices in their classrooms are more likely to understand their students' perspectives and views, resulting in a better learning environment. Since children are already using digital devices entertainment, there should be fewer problems if the teacher considers reintroducing these devices for learning. Additionally, digital devices may also help children respond to classroom tasks and enhance their motivation for learning overall (Sarker et al., 2019).

Alongside language, culture is also learned through one's own experiences and social interactions. Digital storytelling helps teachers to create a bridge between language and culture, allowing students to express their cultural identities through stories. By incorporating digital devices into language learning, teachers can create a more inclusive and dynamic learning environment that caters to the diverse linguistic and cultural backgrounds of their students. It also encourages students to engage with language in a more interactive and meaningful way, as they create their own digital stories that reflect their cultural backgrounds and experiences. These stories can lead to a deeper understanding and appreciation of different languages and cultures, ultimately enhancing the language learning experience for multilingual learners. Digital storytelling has the potential to revolutionize the way language teaching is approached in a culturally sensitive manner. It provides engaging and interactive learning experiences for multilingual learners, allowing them to immerse themselves in diverse cultural narratives and linguistic expressions (Karakuş et al., 2020).

3. Digital storytelling and emerging technologies in SLL

3.1. Elements of cohesive digital storytelling

The major components of digital storytelling are connected like building materials. For instance, audio mix sounds and music to create a mood. The narrator voice illustrates the characters, conflicts and the story. Creating the world and characters of the story helps learners visualize the action. All these elements usually are present in digital storytelling alongside the language components. Despite the common perception, digital storytelling does not need to be limited to computers or other digital technologies. Storytelling remains the key element, defined as the transmission of meaningful messages and moral values. Illustrations of the story do not need to be animated, even though it is quite possible with the current generative AI technologies. Such technologies are also handy for generating background music, or even a narrator voice, but still, a teacher can also be the voice of the story.

The most important element in digital storytelling is its ability to engage students in the learning process. To be effective, a digital story must exhibit internal coherence across all its elements. Firstly, the narrative should be well-structured. For example, it should have a clear beginning and an end, as it is not just a lesson, but a journey for the learners. Alongside its clear structure, it should also have a well-defined conflict, and a resolution for it. This can also be the change that happens throughout the story, a change in the main character, or a change in the world itself. This type of structure should maintain learners' interest in the story, keeping them guessing what is going to happen next, or how will the story end. Secondly, the story should be relevant for the learners to be engaging for them. This involves selecting themes, characters and plots that resonate with their individual characteristics. Such an approach, which considers the personalization of the story, should help learners stay motivated for language learning.

In addition, there could be two creative and expressive ways for multilingual to use digital storytelling in classroom. While adult learners would feel a sense of belonging through sharing their cultural heritage, language learners may have a richer experience in smaller groups through sharing their personal digital designs that express their personal narratives. Language learners can be encouraged and guided in a safe learning environment to tell their own digital stories. Based on their language proficiencies, these creative activities can be guided and structured by the instructor step by step, or they could be more flexible which would allow more self-discovery and creativity for language learners. Providing a safe classroom environment that welcome diversity and positive approaches to learning, students can simply state the world what they want to say (C. Hurlbert, personal communication, 2011).

In creative and innovative shared approaches, students can share their personal stories without being exposed to any judgments in a multicultural learning setting (P. Bizzaro, personal communication, 2016). In a welcoming lesson, multilingual and multicultural learners could share their cultural heritage as new learning opportunities, and this may change the way they perceive the concept of empathy (R. Crane-Bizzaro, personal communication, 2016). Beginner-level language learners, regardless of their age, can even retell a story with a new storyline or ending. Such a flexibility can create new characters in new settings with a familiar plotline. All these innovative and digital approaches would certainly enrich the language learning environment by making the learning more meaningful through activating students' creative skills (D. Hanauer, personal communication, 2011).

Another element to keep in mind is the consistency between the style and tone of the story. Maintaining a consistent style and tone for the illustrations, audio, and the mood of the story helps enhance the storytelling experience for learners. Additionally, effective use of digital materials and media is also essential in terms of enrichment of the story. By utilizing the right applications for desired outcomes, teachers are able to create meaningful and visually pleasing digital experiences for their students. Generative AI tools like ChatGPT can be used to create a base structure, or even the story itself, by providing only simple prompts that describe the elements of the story. For the artwork and illustrations, teachers can use Generative AI tools like Midjourney or DALL·E to make their digital stories colorful and appealing. Applications like Canva, Adobe Express, or even PowerPoint can be used to merge these illustrations with the story to produce the final digital storytelling material. Finally, the inclusion of interactive components, such as quizzes, prompts, or discussion questions can enhance the learning process overall. For example, specific tools like Twine can be used to create interactive stories for students, in which they are the ones to decide how the story advances and ends. There are many digital tools and applications available for teachers to create cohesive and effective digital stories. These assets are not only helpful for teachers in creating teaching materials and content, but also essential for students to be creative and expressive in their target languages (Figure 3.1).

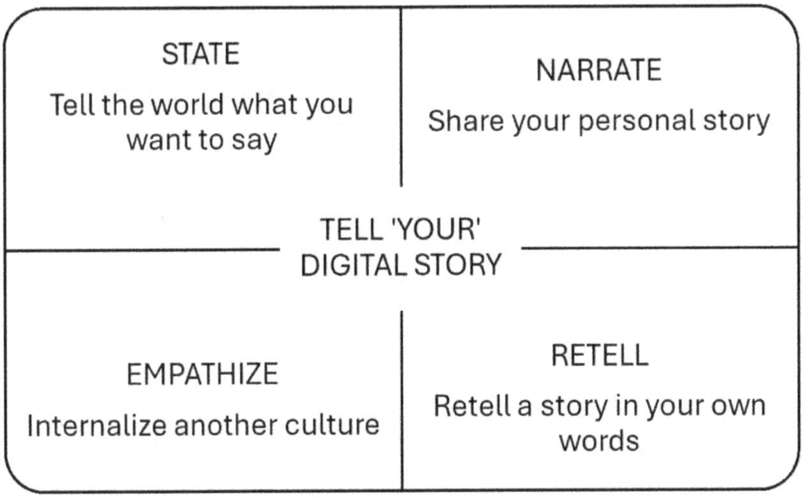

Figure 3.1. Tell your digital story

For any language teacher, the very act of teaching requires not only providing knowledge but also engaging students through the art of storytelling. In the context of second language learning, storytelling is recognized as an important social activity in which students share and learn from narratives. Specifically, storytelling is considered beneficial for expressing thoughts and feelings, for promoting social and emotional growth, while developing a sense of motivation and self-confidence during language learning. Digital native learners with short attention spans, traditional classrooms that emphasize grammar instead of communication, limited time, and inadequate personal guidance are amongst contradicting factors that exist in language learning.

To counter these negatives with the desire to establish a more ideal environment for learners, the digital storytelling approach proposes to aid the language learning process by encouraging the use of the language through fictional, imaginative, or authentic, adding an auditory or visual description to a storyline. To help improve learners' comprehension skills, storytelling techniques can be used while exposing learners to the uncertainties of the language. In this context, the implementation of digital storytelling applications using artificial intelligence (AI) can help create a personalized and adaptive learning experience for language learners. Artificial intelligence can analyze the learner's performance, identify their strengths and weaknesses, and provide customized feedback and learning activities to address their specific needs. This can be particularly beneficial for multilingual learners, as AI systems can recognize their unique language proficiency levels and tailor the storytelling content and exercises to meet their needs. AI-powered digital storytelling can incorporate various media such as poetry, graphic novels, comics, illustrations, and games to enhance the learners' engagement and comprehension of the target language. In the language classroom, a well-composed story can be an opportunity for learners to use a variety of language competences including spelling, grammar, vocabulary, and sentence structure. Moreover, students can increase engagement through the embodied cognitive associations created by rich stories (Maureen et al., 2020). With digital storytelling and emerging technologies, teachers have the ability to craft and design digital materials that provide interactive and engaging lessons. These digital resources allow students to navigate through a virtual world of language rules and principles, making learning a language and its grammar not only comprehensible but also enjoyable.

Moreover, the flexibility of digital storytelling enables the creation of thematic-based story books, overflowing with vibrant illustrations and immersive narratives. These colorful digital materials transport students into captivating worlds, sharpening their listening, speaking, reading, and writing skills in the target language (Yang et al., 2022). Children's books, tales, legends, and storybooks treat the students as the heroes of the story, and through language exposure in this manner, they can integrate linguistic forms into meaningful contexts.

Modern language teaching research on digital storytelling is complemented by other contemporary research areas such as the use of artificial intelligence to support pedagogy in the classroom. The use of AI, without any doubt, brings about new opportunities for both teachers and learners. In second language learning, AI can be used to improve both language development and technological abilities of children. Using digital story writing with AI in classrooms can be considered an effective inquiry-based pedagogical approach to address their language learning. This has the potential to boost learners' problem-solving skills, as well as fostering creativity (Ng et al., 2022). With the assistance of AI, students receive the opportunity to learn at their own pace, have a greater amount of control over their learning, such as having longer time to practice or personalizing their learning process, making them autonomous learners in the process. The success or failure of the teacher, on the other hand, depends on their technological capability and adaptability because the use of AI in teaching definitely requires assistance instead of replacement of the teacher. One example of this is the pedagogy connected with dialogic reading, which uses conversational agents (CAs) in classrooms to engage in dialogical interactions for fostering language development in language learners. In such projects, AI is used to replicate the benefits of dialogic reading with human partners. These interactions with the AI agent can help young language learners develop narrative skills with literacy practice, enhancing narrative-relevant vocalizations and story comprehension overall (Xu et al., 2022).

The emergence of digital technologies is transforming the landscape of language education, leading to the development of innovative methodologies for second and foreign language acquisition. These include computer-assisted language learning (CALL), mobile-assisted language learning (MALL), and game-based language learning (GBLL), among others. Among these evolving methods, digital storytelling has gained increasing recognition as a powerful pedagogical tool. Through digital storytelling, students can engage in creative narrative construction while practicing language skills in a meaningful and personally relevant way.

With the real-world multimedia environment created by digital media, students are now able to explore narratives, communicate with others, and practice language skills in diverse, interactive formats. Digital storytelling enhances this process by allowing learners to craft and share their own stories using multimedia tools, thus promoting both linguistic development and digital literacy. According to Kim et al. (2021), the development of virtual reality (VR) and augmented reality (AR) introduces numerous possibilities for language teaching and learning, particularly in fostering cultural connections. Integrating digital storytelling into VR and AR environments enables learners to become active participants in immersive story-based experiences that simulate real-world cultural contexts. VR, defined as a three-dimensional computer-generated environment that users can explore and interact with, is one of the most promising educational technologies in the realm of language learning. It can place students in entirely new environments where they are required to perform a variety of communicative tasks (Huang et al., 2021). For example, VR and AR can create immersive experiences that not only support language acquisition but also enrich learners' understanding of cultural nuances. When combined with digital storytelling, these technologies allow young multilingual learners to author and navigate culturally rich stories, reinforcing both language and intercultural competence.

VR simulations enhanced by digital storytelling offer a compelling medium through which children can learn language through context-driven, narrative-based interaction. Furthermore, according to Nanduri & Bonsignore (2023), artificial intelligence (AI) can contribute to the preservation of endangered languages and cultures. This approach can be amplified through VR and digital storytelling, establishing early awareness in young learners of linguistic and cultural diversity. By participating in digital storytelling activities, students can engage more deeply with the cultural and linguistic heritage embedded within the language learning process.

By promoting early exposure and appreciation for different languages and traditions, implementation of digital tools such as AI and VR in digital storytelling can contribute to a more inclusive and globalized learning environment and foster a culturally sensitive generation by addressing the need to protect such languages and cultures in a rapidly changing world. These technologies can also facilitate interactive storytelling in different languages, providing a platform for diverse cultural narratives to be shared and celebrated. This innovative approach to language teaching has the potential to greatly enhance culturally sensitive language education for multilingual learners of all ages.

4. Cultural sensitivity in SLL

Traditional language education settings are criticized by linguistic minorities around the world, stressing the issue of the cultural indigeneity of the learners (Nguyen, 2022). The concept of culturally relevant language education responds to these needs, drawing attention to the existing cultural assets in the communities and using them to promote the linguistic and cultural potential of individuals (Allen et al., 2017). The starting point is not the ineffectiveness of the minority groups and their diversity, but the valuable reservoir of varied knowledge and experiences represented by the language, personal, and cultural attributes of the learners, which require representational reinforcement and nurturing. Cultural recognition in educational settings can help learners develop their full social and cultural potential within their communities.

Integrating learners' experiences from their home or other familiar multicultural contexts into early education is an important tool for the positive socialization of children. In daily interaction with young children, it has been observed that using children's familiar culture as a starting point for learning contributes to young children's enthusiasm for oral language practice and language learning in general, since children are proud to announce their individuality and uniqueness loudly (James, 2022). Respect for the linguistic and cultural diversity of young multilingual learners and the integration of their interests into language learning and their use within classroom settings should contribute to the educational achievement of these children.

For culturally sensitive teaching models to be illustrative of the widest variety of learners, they must include the stories of both culturally diverse and local community models that shape the lives of the children. In this context, digital storytelling is promising for sharing authentic events, knowledge and experiences, customs, traditions, and indigenous expressions of language, and thereby can be used to represent the language as a characteristic of the community. Through digital storytelling, young multilingual learners can establish strong ties between stories of real people, meaningful cultural experiences, and linguistic forms that reflect human life and communication. Young multilingual learners can also promote their own languages or the languages of their own choosing. For these children, digital storytelling also has an interactive learning function that allows them to improve phonology and spoken vocabulary (Mohammed Al-Amri, 2020). With the aid of digital storytelling, phonological competence can enhance learners' understanding of language sounds and patterns, leading to more effective communication skills in multiple languages. Additionally, incorporating culturally relevant stories can foster a sense of belonging and pride in learners' diverse linguistic backgrounds. This approach to language teaching aligns with the principles of culturally sensitive and multicultural education and promotes inclusive learning environments for young multilingual learners.

In the context of language teaching, cultural sensitivity can be understood as the willingness to accept learners' perceptions and willingness to take learners' values into consideration in the learning process. This also appeals to adult language learners. Teacher candidates are often challenged by the lack of emotional engagement among students during their language teaching practices, and learners' lack of interest lowers teachers' performance in return (Abu-Rmaileh, 2020). In contrast with children, adult learners are more aware of their learning process which makes them metacognitively aware on what type of teaching approaches increase their learning motivation. One functional pedagogical remedy to increase the sense of belonging is to lower the affective filter through inclusive and culturally sensitive teaching for language learners of all ages (Herrera, 2022).

In this respect, flexibility and openness about the nature of cultural identity and linguistic diversity are necessary qualifications in order to maintain cultural sensitivity. Learners of different ethnic and language backgrounds can experience culture shock in classroom settings, especially when it comes to learning. Culturally diverse learners who are learning a language may feel that they are losing their native culture (Kim, 2020). This is why they may tend to resist activities that involve target language use, since using the target language means embracing and integrating into the culture of the language in question. Therefore, in order to encourage language learning and achieve positive results, teachers should make learners feel they are in an inclusive educational environment in which their cultural identity is valued. Establishing a culturally sensitive environment involves using authentic teaching materials that reflect learner backgrounds and experiences. These materials should be capable of promoting content knowledge and critical thinking while addressing cultural sensitivity and diversity. Learners tend to underachieve when these materials do not reflect their cultural identities by failing to recognize their worth, capability, or potential (Eden, 2024). The choice, creation, and structuring of cultural contexts in language teaching settings are crucial for multilingual learners. For teachers, providing authentic and positive environments that are child-centered naturally requires work in advance and high-level preparation where pedagogical knowledge meets sociocultural and linguistic knowledge. Teachers should not tolerate culture and language-specific deficits in school curricula. A democratic society should invest in a culturally diverse education system for language teaching, transcending stereotypes and neutralizing minority-majority divisions, especially when young children are at stake.

5. Future research directions

Researchers in language education and other education fields need to establish more open-minded approaches for including cultural sensitivity and using digital storytelling resources. In addition, researchers need to pay more attention to how the use of digital storytelling resources actually factors into each of the currently identified elements of culturally sensitive language teaching, particularly for multilingual learners. The affordance of the tools for creating digital storytelling materials alone may prevent the inclusion of some of these elements or may make the inclusion itself too burdensome for some teachers. On the other hand, integrating digital storytelling materials and related digital tools into a language learning ecosystem means that time spent on creating and interpreting is complementary to the overall goals for teachers who are culturally sensitive and aware. This chapter shows how digital storytelling, when implemented within culturally sensitive frameworks for language teaching, makes it possible to meet the linguistic and cultural needs of both young multilingual children from diverse linguistic and cultural backgrounds and young or adult learners overall. In this study, it is proposed that institutions that prioritize translanguaging pedagogies and multiculturalism, and approaches such as digital storytelling, will enable teachers to better implement their teaching methods, for improved motivation, engagement, and language learning outcomes of students in return.

Future research should also focus on training teachers to become critically reflexive about language and to aim for an understanding of what it means to propose a dialogue between languages and cultures. Continued examination of the learning processes brought about by digital storytelling is also essential, as are studies that verify the effects of digital storytelling on the learning and motivation of students with different learning styles and who are at different stages of language proficiency. Additionally, the possible effects of digital storytelling on language learning and its assessment should also be explored. Finally, the familiarity of teachers with digital tools should also be explored to address the prevalence of using digital storytelling, particularly over the long term or in different teaching and learning contexts.

6. Conclusion

By learning through digital narratives, multilingual learners may potentially develop a more positive and stronger connection to their heritage language, the linguistic diversity of their societies, and the cultures from around the world, which could also significantly benefit their literacy and cognitive development, self-esteem, identity, and inclusive social integration (Ortega, 2020). These findings suggest that culturally and linguistically sensitive language teaching practices focus on diverse community values, languages, and collective narratives could effectively contribute to multilingual literacy education in diverse multilingual classroom settings. In today's multicultural and interconnected world, it is important to appreciate, learn from, and respect all cultural and linguistic resources, repertoires, and practices in diverse multilingual educational settings. In addition, these case studies also provide many innovative and useful ideas and suggestions for language learners' personalized digital story design, culturally sensitive digital storytelling and literacy practices, and associated digital pedagogy.

The recent developments in digital tools, apps, and platforms facilitate the provision of responsive, flexible, diverse, and inclusive teaching models of language education, especially for language learners and those from multilingual, culturally, and cognitively diverse communities. The findings of this study show that digital tools incorporating digital storytelling resources, multimedia, and interactive features could generate rich, dynamic, student-centered, culturally and linguistically appropriate learning materials and experiences, which could effectively engage and support multilingual learners. It is clear that digital storytelling could play a highly significant role in enhancing young multilingual students' second language development, literacy practices, and learning motivation through more creative, social, and communicative activities, and a positive identity with their cultures and languages.

References

Abu-Rmaileh, S. (2020). Teacher stress, its effects and management. In C. Coombe, N. J. Anderson, & L. Stephenson (Eds.). *Professionalizing your English language teaching* (pp. 39-52). Springer International. https://doi.org/10.1007/978-3-030-34762-8

Allen, A., Hancock, S. D., Starker-Glass, T., & Lewis, C. W. (2017). Mapping culturally relevant pedagogy into teacher education programs: A critical framework. *Teachers College Record, 119*(1), 1-26. https://doi.org/10.1177/016146811711900107

Anderson, J., & Macleroy, V. (2020). Connecting worlds: Interculturality, identity and multilingual digital stories in the making. In P. Holmes & J. Andrews (Eds.). *Education and migration* (pp. 126-149). Routledge. https://doi.org/10.4324/9780429057878

Androutsopoulos, J. (2021). Investigating digital language/media practices, awareness, and pedagogy: Introduction. *Linguistics and Education, 62*, 1-4. https://doi.org/10.1016/j.linged.2020.100872

Eden, C. A., Chisom, O. N., & Adeniyi, I. S. (2024). Cultural competence in education: Strategies for fostering inclusivity and diversity awareness. *International Journal of Applied Research in Social Sciences, 6*(3), 383-392. https://doi.org/10.51594/ijarss.v6i3.895

Guillén, G., Sawin, T., & Avineri, N. (2020). Zooming out of the crisis: Language and human collaboration. *Foreign Language Annals, 53*(2), 320-328. https://doi.org/10.1111/flan.12459

Hector, E. (2022). Mind, motivation, and meaningful learning: Strategies for teaching adult learners, by Melissa L. Miller. *Canadian Journal of Academic Librarianship, 8*, 1-3. https://cjal.ca/index.php/capal/article/view/39167

Herrera, S. G. (2022). *Biography-driven culturally responsive teaching: Honoring race, ethnicity, and personal history.* Teacher's College Press.

Huang, X., Zou, D., Cheng, G., & Xie, H. (2021). A systematic review of AR and VR enhanced language learning. *Sustainability, 13*(9), 4639. https://doi.org/10.3390/su13094639

James, A. (2022). Talking of children and youth: Language, socialization and culture. In V. Amit & H. Wuff (Eds.), *Youth cultures* (pp. 43-62). Routledge.

Karakuş, M., Türkkan, B. T., & Namlı, N. A. (2020). Investigation of the effect of digital storytelling on cultural awareness and creative thinking. *Education & Science/Egitim ve Bilim, 45*(203). https://doi.org/10.15390/EB.2020.8576

Kendrick, M., Early, M., Michalovich, A., & Mangat, M. (2022). Digital storytelling with youth from refugee backgrounds: Possibilities for language and digital literacy learning. *TESOL Quarterly, 56*(3), 961-984. https://doi.org/10.1002/tesq.3146

Kim, D., Coenraad, M., & Park, H. R. (2021). Digital storytelling as a tool for reflection in virtual reality projects. *Journal of Curriculum Studies Research, 3*(1), 101-121. https://doi.org/10.46303/jcsr.2021.9

Kim, J. (2020). International students' intercultural sensitivity in their academic socialisation to a non-English-speaking higher education: a Korean case study. *Journal of Further and Higher Education, 44*(7), 939-955. https://doi.org/10.1080/0309877X.2019.1627298

Kincl, T., & Štrach, P. (2021). Born digital: Is there going to be a new culture of digital natives? *Journal of Global Scholars of Marketing Science, 31*(1), 30-48. https://doi.org/10.1080/21639159.2020.1808811

Kristiawan, D., Ferdiansyah, S., & Picard, M. (2022). Promoting vocabulary building, learning motivation, and cultural identity representation through digital storytelling for young Indonesian learners of English as a foreign language. *Iranian Journal of Language Teaching Research, 10*(1), 19-36. 10.30466/ijltr.2022.121120

Liang, Q., de la Torre, J., & Law, N. (2021). Do background characteristics matter in children's mastery of digital literacy? A cognitive diagnosis model analysis. *Computers in Human Behavior, 122*, 106850. https://doi.org/10.1016/j.chb.2021.106850

Maureen, I. Y., van der Meij, H., & de Jong, T. (2020). Enhancing storytelling activities to support early (digital) literacy development in early childhood education. *International Journal of Early Childhood, 52*(1), 55-76. https://doi.org/10.1007/s13158-020-00263-7

McGee, P. (2014). *The instructional value of digital storytelling: Higher education, professional, and adult learning settings*. Routledge. https://doi.org/10.4324/9780203066140

Mohammed Al-Amri, H. (2020). Digital storytelling as a communicative language teaching-based method in EFL classrooms. *Arab World English Journal* (AWEJ), *11*(1), 270-281. https://dx.doi.org/10.24093/awej/vol11no1.20

Nanduri, D. K., & Bonsignore, E. M. (2023). *Revitalizing endangered languages: AI-powered language learning as a catalyst for language appreciation*. ARXIV Cornell University. https://doi.org/10.48550/ARXIV.2304.09394

Ng, D. T. K., Luo, W., Chan, H. M. Y., & Chu, S. K. W. (2022). Using digital story writing as a pedagogy to develop AI literacy among primary students. *Computers and Education: Artificial Intelligence, 3*, 100054. https://doi.org/10.1016/j.caeai.2022.100054

Nguyen, T. T. T. (2022). Educational linguicism: linguistic discrimination against minority students in Vietnamese mainstream schools. *Language Policy, 21*(2), 167-194. https://doi.org/10.1007/s10993-021-09601-4

Office of Communications. (2023). *Children's media use and attitudes report 2023.* https://www.ofcom.org.uk/media-use-and-attitudes/media-habits-children/children-and-parents-media-use-and-attitudes-report-2023/

Ortega, L. (2020). The study of heritage language development from a bilingualism and social justice perspective. *Language Learning, 70,* 15-53. https://doi.org/10.1111/lang.12347

Rahiem, M. D. (2021). Storytelling in early childhood education: Time to go digital. *International Journal of Child Care and Education Policy, 15*(1), 4. https://doi.org/10.1186/s40723-021-00081-x

Robin, B. R. (2008). Digital storytelling: A powerful technology tool for the 21st century classroom. *Theory into Practice, 47*(3), 220-228. https://doi.org/10.1080/00405840802153916

Sarker, M. N. I., Wu, M., Cao, Q., Alam, G. M., & Li, D. (2019). Leveraging digital technology for better learning and education: A systematic literature review. *International Journal of Information and Education Technology, 9*(7), 453-461. 10.18178/ijiet.2019.9.7.1246

Xu, Y., Aubele, J., Vigil, V., Bustamante, A. S., Kim, Y. S., & Warschauer, M. (2022). Dialogue with a conversational agent promotes children's story comprehension via enhancing engagement. *Child Development, 93*(2), e149-e167. https://doi.org/10.1111/cdev.13708

Yang, Y. T. C., Chen, Y. C., & Hung, H. T. (2022). Digital storytelling as an interdisciplinary project to improve students' English speaking and creative thinking. *Computer Assisted Language Learning, 35*(4), 840-862. https://doi.org/10.1080/09588221.2020.1750431

Chapter 4

Empowering Chinese language teachers through AIGC and digital literacy

Jiawen Zhu
East China Normal University

Fang Cao
Seoul National University

Xiaoqing Gu
East China Normal University

Abstract

This chapter explores the potential of AI-Generated Content (AIGC) in Chinese language education. It first outlines challenges in traditional approaches—pedagogical rigidity, low engagement, limited feedback, and scalability—before analyzing how AIGC addresses them through personalized input, interactive output, and culturally adaptive resources. Grounded in second language acquisition (SLA) theories, the chapter situates AIGC as both a pedagogical and technological innovation. A case study of a phased seminar demonstrates short-term gains in digital literacy and resource creation, alongside long-term implications for teacher training and ethical practice. The chapter concludes by emphasizing workflow-based integration and global applicability in shaping future language education.

Keywords: Language education, Chinese language, Artificial intelligence, AIGC, digital literacy

* * *

1. Introduction

Amid an unprecedented technological revolution, educators increasingly recognize artificial intelligence (AI) as a driving force transforming educational landscapes worldwide. AI research focuses on developing machines capable of mimicking human cognitive processes, aiming to provide users with experiences similar to interacting with real humans (Sharadgah & Sa'di, 2022). This technology has been particularly influential in the field of language learning, where AI-driven tools, such as natural language processing (NLP) systems, have significantly advanced computer-assisted language learning (CALL; Almelhes, 2023). For instance, adaptive learning technologies and AI have been recognized as significant advancements in educational technology, providing novel pathways for personalized and efficient learning experiences (Becker et al., 2018). Moreover, learners often report feeling more confident when learning languages through AI-driven chatbots compared to human tutors, underscoring the potential of AI to create supportive and effective learning environments (Haristiani, 2019).

Building on these advancements, AI-generated content (AIGC) has emerged as a transformative technology in the education sector. AIGC refers to the generation of relevant and context-specific content through artificial intelligence techniques, such as generative adversarial networks and large pre-trained models, which learn and generalize from existing data to create novel materials (Lee et al., 2020). Unlike traditional AI applications that primarily analyze data, AIGC generates diverse resources, including instructional materials, assessments, and interactive tasks tailored to specific needs (Lim et al., 2023; Lu et al., 2023). Scholars have highlighted its potential to empower education by providing scalable and culturally relevant resources, improving learning experiences, and enhancing overall educational effectiveness, while addressing critical challenges such as resource creation and adaptation to varying pedagogical contexts (Kasneci et al., 2023). This capability enables scalable solutions to longstanding issues in language education, such as the development of personalized resources and the facilitation of cultural relevance, emphasizing AIGC's significance in transforming educational practices (Cao et al., 2023; Yang, Yang, & Tong, 2023).

In the context of second language acquisition (SLA), AIGC aligns with foundational theories, such as Krashen's input hypothesis and Swain's (1985) output hypothesis. Krashen's (1982) input hypothesis emphasizes the need for comprehensible input that is slightly beyond the learner's current proficiency level (i + 1) (Chen et al., 2024a), while Swain's (1985) Output Hypothesis highlights the cognitive and metalinguistic benefits of producing language. Rather than treating these theories as abstract principles, this chapter employs them as analytical methods to interpret the affordances of AIGC. Specifically, AIGC can operationalize i + 1 input through adaptive text generation and immersive scenarios, stimulating output by prompting learners to produce creative responses, reflections, or multimodal projects. In doing so, SLA theories not only contextualize the challenges of learning the Chinese language but also help explain why AIGC is well-positioned to address them effectively.

Despite its promise, the integration of AIGC into language education underscores the critical need for teachers to develop robust digital literacy skills. As these tools become more prevalent, educators must understand their functionalities and critically evaluate their applications and limitations (Yang et al., 2023). Effective integration requires teachers to navigate complex technological environments, adapt resources to meet the diverse needs of learners, and address ethical considerations such as bias and over-reliance on technology. Therefore, enhancing educators' digital literacy is crucial for maximizing the benefits of AIGC while mitigating potential challenges (Cao et al., 2023).

This chapter explores the transformative potential of AIGC in Chinese language education, addressing the unique challenges posed by the tonal and logographic nature of the Chinese language. By focusing on the application of AIGC, the chapter examines its role in overcoming limitations of traditional teaching methods and illustrates its adaptability to diverse learner contexts. Subsequent sections delve into case studies demonstrating AIGC's practical applications, as well as discussions on ethical and pedagogical considerations. By equipping educators with the necessary skills to leverage AIGC, this chapter contributes to advancing Chinese language education and preparing learners for success in a digitally interconnected world.

2. Challenges in traditional language teaching and the role of AIGC

Traditional language teaching, often characterized by textbook-driven instruction, grammar-translation methods, rote memorization, and teacher-centered classrooms, has long served as the foundation for language acquisition worldwide. However, as the demands and complexities of global communication increase, educators face growing challenges in delivering effective instruction. In teaching Chinese as a foreign language, these challenges become particularly pronounced due to the language's unique linguistic and cultural characteristics. This chapter identifies key limitations in traditional Chinese language instruction and illustrates how AIGC can help bridge these gaps, while also discussing implementation challenges and considerations.

2.1. Pedagogical constraints and personalized learning

Traditional pedagogical models often lack adaptability to diverse learner profiles (Arslan, 2024). Uniform pacing, static content, and reliance on standardized assessments hinder individualized learning. These learning characteristics are particularly problematic in Chinese language instruction, where learners must master abstract character forms, tonal patterns, and sociocultural nuances. Developing and adjusting instructional materials manually not only adds to teacher workload but also limits scalability.

AIGC has the potential to transform language education by offering personalized learning experiences tailored to each learner's needs (Wang, Wu, et al., 2024). By analyzing data on learner progress, strengths, and weaknesses, AIGC can generate customized materials and exercises. Such personalization ensures that learners encounter comprehensible input slightly beyond their current proficiency level, resonating with Krashen's input hypothesis (Krashen, 1982). By providing targeted opportunities for practice, AIGC enables learners to build confidence and achieve mastery in specific areas of the language. Nevertheless, the effectiveness of these experiences depends on educators' digital literacy, particularly their ability to interpret AI-generated insights and apply them effectively in classroom practice.

2.2. Learner engagement and interactive experiences

Student engagement and motivation are critical factors in the success of language learning. However, traditional language classrooms often struggle to maintain high levels of student engagement. The reliance on repetitive drills, limited interactivity, and lack of authentic application can lead to disengagement and diminished motivation (Dashtestani, 2014).

For Chinese language learners, mastering characters, tones, and grammar rules can be daunting. Without engaging and meaningful learning experiences, students may find the process tedious and overwhelming, leading to high dropout rates and poor learning outcomes. This challenge is further compounded for Western learners, who often perceive Chinese tones and logographic characters as exceptionally difficult (Shen, 2018). Notably, the U.S. Department of State (n.d.) classifies Mandarin and Cantonese as "super-hard languages," requiring over 2,200 hours of instruction to achieve professional proficiency—significantly more than the 600 hours needed for Spanish. This disparity highlights the cognitive and emotional demands placed on learners, which can undermine their expectations of success (Hu, 2010).

AIGC has the potential to enhance student engagement by creating interactive and immersive learning experiences (Chiu, 2024; Diwan et al., 2023). Through the use of simulations, virtual reality, and conversational agents, AI can bring language learning to life in ways that traditional methods cannot (Divekar et al., 2022). These tasks encourage learners to produce language, identify gaps, and refine their knowledge—processes central to Swain's (1985) output hypothesis. By situating practice in realistic contexts, AIGC not only improves proficiency but also strengthens cultural competence, though effective integration depends on teachers' readiness to use these tools strategically.

2.3. Assessment and feedback

Accurate assessment and timely feedback are essential components of effective language teaching (Seker & Dincer, 2014). However, traditional methods of assessment, such as written exams and oral tests, often fall short in capturing the full spectrum of a student's language abilities. These methods tend to focus on specific aspects of language proficiency, such as grammar and vocabulary, while neglecting other critical components like cultural understanding and communicative competence. In large classrooms, providing timely, individualized feedback is especially difficult (Çağlayan, 2024).

AIGC can enhance this process by delivering more accurate and immediate feedback (Wang, Wang, Xu, et al., 2023). AI-powered assessment tools can analyze a student's spoken and written language, identifying errors and offering constructive feedback in real-time. This immediate feedback allows students to correct mistakes and reinforce learning on the spot, leading to more effective language acquisition.

For Chinese language learners, AIGC can evaluate pronunciation accuracy, tone recognition, and character writing, offering detailed feedback that traditional methods may overlook. Automating these processes not only reduces teacher workload but also allows educators to devote more attention to targeted support. Yet, effective use of AI-generated feedback requires educators to have sufficient digital literacy to interpret results and provide complementary human guidance.

2.4. Accessibility and scalability in language education

Traditional language education depends heavily on qualified teachers and in-person delivery, limiting reach in remote or under-resourced areas. This is particularly problematic for language instruction, where the supply of skilled educators cannot meet the demand (Li & Siu, 2009). Access to quality resources is also uneven, especially in rural or low-income areas. Traditional approaches may not adapt well to diverse contexts, exacerbating educational disparities. Teachers with limited digital literacy may further struggle to leverage online platforms effectively (Maja, 2023).

AIGC expands access by supporting distance education and automating resource development. AI-generated content, ranging from vocabulary exercises to culturally adapted dialogues, can be disseminated widely and tailored to local learner contexts. AIGC can support large-scale language education programs by automating administrative tasks and providing scalable solutions for content delivery and assessment (Guo et al., 2023). To realize these benefits, however, educators must possess the digital literacy required to manage and implement AI tools effectively.

2.5. Considerations and challenges in using AIGC

2.5.1. Ethical considerations

While AIGC offers significant benefits, it raises ethical concerns. Data privacy is a primary issue, as these tools often rely on large amounts of learner data to generate personalized content (Cao et al., 2023). Compliance with privacy regulations and safeguarding personal information are essential. Another concern is algorithmic bias, as AI models trained on large datasets may reproduce cultural stereotypes or inaccuracies (Guo et al., 2023). Educators must be vigilant in reviewing AIGC materials to ensure that they are culturally appropriate and free from bias. This task requires a high level of digital literacy, as educators must be able to critically assess AI-generated content and make informed decisions about its use in the classroom.

2.5.2. Teacher training and readiness

Successful integration of AIGC depends on adequately prepared educators (Moorhouse & Kohnke, 2024; Tafazoli, 2024). Beyond technical skills, teachers require pedagogical strategies for embedding AI meaningfully in curricula (Al-Zyoud, 2020). Resistance to change remains a barrier, as some educators hesitate to adopt AI or lack confidence in its use (Kohnke et al., 2023; Nazaretsky et al., 2022). Continuous professional development and peer support are essential. Digital literacy should remain central, enabling teachers to interpret AI insights, adapt them for context, and balance technological possibilities with pedagogical integrity.

2.5.3. Dependence on technology

Although AIGC enhances efficiency, over-reliance risks diminishing essential human aspects of teaching such as empathy, cultural understanding, and interpersonal communication (Chen et al., 2024b). Language learning is relational, and teachers must balance AI use with human interaction. Technical issues, including connectivity problems, system failures, and hardware limitations, also pose risks. Educators need contingency strategies to manage such disruptions and ensure continuity of learning. Strong digital literacy again plays a crucial role, enabling teachers to troubleshoot issues and integrate AIGC without compromising instructional quality.

2.5.4. Quality control

Ensuring the quality and reliability of AI-generated content remains a critical challenge. Although AIGC can generate vast amounts of material, not all outputs are accurate, pedagogically sound, or culturally sensitive. Human oversight is crucial for reviewing, curating, and refining these resources (Lim et al., 2023). In Chinese language education, this includes verifying the accuracy of AI-generated translations, character writing exercises, and pronunciation guides. Without such a review, AIGC risks reinforcing stereotypes or introducing errors. Educators' digital literacy is central to this process, equipping them to critically evaluate AI outputs and adapt them to meet the needs of their students.

3. Case study: integrating AIGC into international Chinese education through short video production

3.1. Case background

In recent years, the field of international Chinese language education has faced numerous challenges, particularly in adapting traditional teaching methods to meet the diverse needs of a global student population. The complexity of the Chinese language, characterized by its logographic script, tonal pronunciation, and deep cultural roots, often presents significant learning barriers, especially for non-native speakers. Moreover, the traditional approach to teaching Chinese, which heavily relies on rote memorization and standardized materials (Yang & Gao, 2020), has struggled to keep pace with the dynamic and interactive learning environments expected by today's students (Wang, Wang, & Zou, 2024). To address these challenges, a new practical seminar titled "AIGC + International Chinese Education Short Video Production" was developed for master's students specializing in International Chinese Education. This seminar aimed to explore innovative methods of teaching resource development through the integration of AIGC, focusing on the creation of diverse and culturally rich short videos that would enhance both teaching and learning experiences.

Before the implementation of AIGC, educators faced a substantial burden in developing teaching materials that were not only engaging and culturally appropriate but also tailored to the varying linguistic levels and cultural backgrounds of their students. Traditional content creation processes were time-consuming and often resulted in materials that lacked the interactivity and relevance needed to maintain student engagement (Liu et al., 2023). Additionally, educators often found traditional learning materials to be monotonous and disconnected from real-world applications, leading to decreased motivation and participation in language learning. The limitations of available resources, compounded by the lack of digital literacy among both students and educators, further hindered the effective use of technology in enhancing educational outcomes. These challenges underscored the need for a more dynamic and responsive approach to content creation, which could be addressed through the integration of AIGC.

3.2. Implementation

The "AIGC + International Chinese Education Short Video Production" seminar was carefully structured to integrate AIGC tools into the content creation process in a systematic and pedagogically sound manner. The course began with an in-depth orientation session, where educators and pre-service educators were introduced to the various AIGC tools available for educational use. This session was designed not only to familiarize participants with the technical aspects of the tools but also to highlight their potential applications in creating engaging and contextually relevant educational content.

3.2.1. Phase one: cognition and attitudes toward AIGC

The first phase of the seminar, titled "Cognition and Attitudes toward AIGC," aimed to introduce educators to the fundamental principles of AI and the development of AIGC tools. This phase provided a comprehensive overview of AI's basic concepts, including how AI technologies can mimic human cognitive processes and generate content autonomously. The session also focused on the role of AIGC in international Chinese language teaching, emphasizing its potential to assist educators in digitizing teaching materials and combining these resources to create more interactive and engaging video lessons.

A key component of this phase was cultivating digital literacy among educators. Educators were not only introduced to the technical aspects of AI but also encouraged to critically assess how these tools could be applied in educational contexts. Discussions centered on the ethical implications of AI, data privacy concerns, and the importance of using AI tools responsibly. By the end of this phase, educators had developed a foundational understanding of AI and AIGC, along with an awareness of the critical role that digital literacy plays in effectively leveraging these technologies in the classroom.

3.2.2. Phase two: selection and application of tools

The second phase, "Selection and Application of Tools," focused on the practical aspects of AIGC by introducing educators to various tools for generating different types of educational content. The phase was structured around four key areas: text generation, image generation, audio generation, and video generation. Each category was accompanied by demonstrations and hands-on practice sessions where educators could explore the tools' functionalities.

This phase was instrumental in developing digital literacy, as educators were guided through the process of selecting and applying the appropriate AIGC tools based on their specific educational needs. For text generation, tools like ChatGPT were highlighted for their ability to produce dialogues, narrations, and explanatory content. Educators practiced crafting effective prompts and critically evaluating the generated content for accuracy and cultural appropriateness.

In the realm of image generation, educators learned to create visual elements like illustrations and animations. Here, digital literacy was further enhanced as educators assessed the strengths and weaknesses of these tools, considering factors such as image quality, cultural relevance, and the potential biases in AI-generated visuals.

For audio and video generation, educators explored tools that could produce voiceovers, background music, and multimedia presentations. This phase emphasized the importance of digital literacy in navigating these tools, allowing educators to make informed decisions about how best to integrate AI-generated content into their educational videos. By the end of this phase, educators had developed a deeper understanding of how to select and apply AIGC tools effectively, a key aspect of digital literacy.

3.2.3. Phase three: content design and production

The third phase, "Content Design and Production," involved the actual creation of the short videos. Building on the knowledge and skills acquired in the previous phases, educators moved forward with conceptualizing, scripting, and producing their educational videos. This phase required educators to apply both their linguistic and technical skills in a creative and meaningful way, further enhancing their digital literacy.

During this phase, educators integrated various forms of AI-generated content—text, images, and audio—into cohesive video presentations. Educators chose topics such as "Chinese Characters Teaching," "Idioms Teaching," and "Classical Poetry Teaching" (Table 4.1). The process involved critical engagement with the tools, where educators needed to modify and adapt AI-generated content to meet the educational objectives of their projects. This hands-on experience was crucial for developing the technical competencies required to use digital tools effectively in educational settings.

Moreover, this phase encouraged educators to reflect on the broader implications of using AI in education. They considered how AIGC could enhance or hinder the learning experience and ways AI might complement, rather than replace, the human elements of teaching. This reflection deepened their digital literacy, as it required them to think critically about the role of technology in education and their own responsibilities as educators.

Instructional Topics	Related Image	AIGC-Enable Content Creation
Chinese characters		Text Image Video Audio, and speech generation
Idioms		Text Image Audio, and Digital human generation
Classical poetry		Text Image Video Audio Speech, and digital human generation

Table 4.1. AIGC-enabled content creation for different teaching scenarios in Chinese language education

3.2.4. Phase four: reflection and evaluation

The evaluation of this course is framed around Kirkpatrick's four-level model (1996), systematically examining participants' satisfaction, learning outcomes, behavioral changes, and overall results. This approach provides a comprehensive understanding of the program's effectiveness in integrating AIGC tools into international Chinese language teaching.

First, at the reaction level, participants' satisfaction was assessed in four key areas: course planning, operations, support, and outcomes. Overall, feedback was highly positive, with over 90% of participants indicating that the course met their expectations. They particularly appreciated the strong theoretical foundation and the practical sessions, which allowed them to explore the integration of AIGC tools into classroom teaching. Many participants noted that the course expanded their understanding of digital technology and its potential in education, further reinforcing their confidence in adopting AI tools for pedagogical purposes.

Second, at the learning level, to evaluate changes in participants' knowledge, skills, and attitudes, experts assessed their AIGC-generated teaching materials across six dimensions: tool mastery, teaching applicability, content completeness, visual quality, content accuracy, and creativity. The analysis revealed that most participants successfully designed and produced digital teaching materials that aligned with instructional goals while addressing cultural and linguistic nuances. However, the dimension of "creativity" scored relatively lower, suggesting that future iterations of the course should place greater emphasis on fostering originality and encouraging participants to develop innovative, personalized educational materials instead of replicating predefined examples.

Next, at the behavior level, semi-structured interviews provided insights into participants' behavioral changes and their intentions to apply AIGC in practice. The interviews revealed a generally positive shift in participants' attitudes and skillsets. Participants noted that the course enhanced their perspectives on AIGC's role in international Chinese language teaching, with many citing improved efficiency in lesson preparation and enhanced teaching capabilities. However, some participants emphasized the need for manual adjustments to AIGC-generated content, viewing these tools as supplementary rather than standalone solutions. Participants also expressed strong interest in more systematic and in-depth training programs, coupled with the creation of a digital resource-sharing platform for international Chinese language teachers.

Additionally, in results level, the ultimate effectiveness of the program was measured through online pre- and post-training assessments of participants' AIGC literacy. The results demonstrated notable progress in participants' understanding and attitudes toward AIGC, as well as their ability to select and apply appropriate AIGC tools. However, improvements were less evident in areas such as designing and creating AIGC content, applying these tools in practice, and conducting evaluations and reflections. These findings suggest that while the training effectively enhanced participants' foundational knowledge and practical selection skills, more extensive support and longer-term practice may be required to fully develop their competencies in advanced application and critical evaluation.

Taken together, the seminar produced clear short-term outcomes, including increased confidence, immediate skill acquisition, and the production of concrete AIGC-based teaching resources. At the same time, the evaluation points to important long-term implications for teacher education: the need for sustained training programs, platforms for resource sharing, and extended practice opportunities to ensure that digital literacy and creativity become embedded competencies within international Chinese language teacher development.

3.3. Reflections

The integration of AIGC in this seminar provided several key insights, particularly in fostering collaborative intelligence and advancing innovative teaching practices. A significant takeaway was the synergistic relationship between human creativity and AI capabilities, emphasizing the educator's critical role in guiding AI to generate culturally and pedagogically relevant content. By crafting precise prompts, educators ensure that AI-generated materials align with specific instructional needs, thereby optimizing the interplay between human expertise and AI-generated options. This iterative process not only broadens the range of available teaching resources but also stimulates innovative approaches that transcend traditional methods.

When viewed in a broader international context, this workflow-based model also resonates with comparable AIGC interventions in teacher education worldwide, such as AI-supported writing feedback in English language teaching or chatbot-mediated conversation practice in European universities. Yet, unlike many AIGC training programs that focus on individual tools or teach functions in isolation, this seminar emphasized the integration of multiple modalities into a complete production workflow. Participants learned to connect text, image, audio, and video generation step by step, culminating in a coherent teaching product. This approach enhanced tool familiarity and fostered transferable skills in designing end-to-end educational resources, marking a key distinction from other AIGC-based interventions. This systemic design highlights the program's potential to serve as a replicable framework for global teacher training initiatives.

Beyond technical skills, the seminar also cultivated new ways of thinking that are integral to digital literacy. By working through a complete production workflow, linking text, image, audio, and video, participants developed systemic thinking and process-oriented problem-solving, learning to view AIGC as part of an integrated task chain rather than isolated tools. Their engagement in critically selecting and adapting AI outputs fostered evaluative and integrative thinking, while the requirement to produce original short videos encouraged creative expression. These experiences demonstrate that digital literacy is not only about operating tools, but also about cultivating critical, integrative, and creative mindsets essential for teaching in the AIGC era.

The seminar also highlighted a shift in pedagogical paradigms, moving from providing pre-made resources to empowering educators with the skills to create personalized materials. This approach, described as "teaching to fish," equips teachers with the ability to localize, customize, and personalize educational content using AIGC tools. This shift addresses persistent challenges in resource creation and fosters a sustainable model for educational innovation. Through training, international Chinese language educators gained the ability to design lesson plans, produce multimedia resources, and adapt materials to diverse cultural and linguistic contexts, thereby creating a dynamic ecosystem for resource development.

Beyond the immediate context of international Chinese language education, the adaptability and scalability of the AIGC training model were underscored through its applications in other educational settings. For instance, AIGC was successfully integrated into university-industry collaborative courses, enabling students to produce digital resources on varied topics. Public elective courses utilized AIGC to develop digital teaching materials for rural primary and secondary schools in China, showcasing its potential in addressing educational inequalities. Additionally, partnerships with STEM programs introduced AIGC in introductory courses, where first-year students created digital content to illustrate foundational scientific principles. These examples illustrate the versatility of AIGC, demonstrating its applicability across disciplines and educational stages.

The seminar also emphasized the need for specialized platforms tailored to the unique requirements of international Chinese education. While general-purpose AIGC tools are valuable, their effective application requires significant expertise, digital literacy, and interdisciplinary knowledge. Developing an integrated platform that combines optimized AIGC tools with user-friendly interfaces, prompt libraries, and instructional tutorials could streamline content creation and enhance accessibility. Furthermore, fostering a community of practice for sharing case studies and best practices would enable educators worldwide to collaborate and innovate, amplifying the impact of AIGC in education.

From an SLA perspective, these reflections also show how AIGC can facilitate both comprehensible input, as learners engage with adaptive resources, and meaningful output, as they create and refine content (Wang, Wang, & Zou, 2024). This dual alignment with input and output hypotheses underscores why AIGC is not merely a technical tool but also a pedagogically grounded innovation. At the same time, the relatively lower evaluations of creativity highlight an important area for further analysis: while AIGC supports efficiency and personalization, genuine creativity emerged when participants went beyond replicating AI suggestions and used the tools as springboards for original, culturally grounded expression.

These reflections underscore the transformative potential of AIGC in enhancing international Chinese language education and provide a blueprint for adapting these insights to broader educational contexts. As human-AI collaboration continues to evolve, thoughtful integration of these tools can enrich the educational experience, addressing diverse learner needs while preparing educators and students for the challenges of a digitally interconnected world.

4. Conclusion: shaping the future of Chinese language teaching

As we reflect on this book chapter, it is clear that AIGC has great potential for international Chinese language education. The integration of AIGC into teaching practices, as exemplified by the "AIGC + International Chinese Education Short Video Production" seminar, demonstrates how AI can address many of the longstanding challenges in language education. AIGC allows educators to create engaging, culturally relevant, and personalized materials. This reduces the workload for teachers while improving learning outcomes. From the perspective of SLA, these outcomes can also be understood as AIGC's ability to provide learners with comprehensible input that is appropriately challenging (Krashen, 1982) and to create opportunities for meaningful output that fosters deeper reflection and knowledge consolidation (Swain, 1985).

However, realizing the full potential of AIGC requires more than just technological integration; it demands a concerted effort in teacher training and professional development. As AIGC continues to develop, it is crucial that educators are equipped with the necessary skills to use these tools effectively and ethically. Teacher training programs should therefore focus on improving digital literacy, providing educators with a deep understanding of how AI technologies work, and offering strategies for integrating them into their pedagogical practices. Moreover, as the ethical considerations surrounding AI become increasingly complex, educators must be trained to understand these challenges, ensuring that the use of AIGC aligns with educational values and student well-being.

To support the effective adoption of AIGC in classrooms, international Chinese language teachers should take proactive steps to enhance their digital literacy. This includes familiarizing themselves with a variety of AIGC tools, experimenting with these technologies to understand their capabilities and limitations, and developing strategies to integrate AI into lesson plans meaningfully. Teachers should also engage with ongoing professional development opportunities, participate in communities of practice where they can share experiences and best practices, and stay informed about the latest advancements in AI and education. By doing so, they can not only improve their own teaching practices but also contribute to the broader discourse on AI in education.

Looking to the future, the landscape of language education is poised for further transformation as AIGC technologies continue to advance. We can anticipate the development of more sophisticated language models that offer even greater levels of personalization, enabling tailored learning experiences that cater to individual educators needs. Additionally, the integration of AI with emerging technologies such as virtual reality could lead to the creation of immersive language learning environments, where educators can practice their skills in realistic, interactive settings. These trends suggest a future where language education is more effective and more engaging and accessible to learners worldwide.

In conclusion, while the journey of integrating AIGC into Chinese language education is still in its early stages, the potential benefits are clear. By embracing these technologies, educators can enhance their teaching practices, provide educators with richer learning experiences, and prepare them for a future where digital literacy and AI proficiency will be essential. As we move forward, it is essential to continue exploring, experimenting, and refining the use of AIGC in education, ensuring that these tools are used to empower both teachers and students, and ultimately, to advance the field of international Chinese language education.

References

Almelhes, S. A. (2023). A review of artificial intelligence adoption in second-language learning. *Theory and Practice in Language Studies, 13*(5), 1259-1269. https://doi.org/10.17507/tpls.1305.21

Al-Zyoud, H. M. M. (2020). The role of artificial intelligence in teacher professional development. *Universal Journal of Educational Research, 8*(11B), 6263-6272. https://doi.org/10.13189/ujer.2020.082265

Arslan, K. (2024). Teaching English vocabulary: Innovative methods. *Contemporary Research in Language and Linguistics, 2*(1), 45-53. https://doi.org/10.62601/crll.v2i1.26

Becker, S. A., Brown, M., Dahlstrom, E., Davis, A., DePaul, K., Diaz, V., & Pomerantz, J. (2018). *NMC horizon report: 2018 higher education edition.* Educause. https://library.educause.edu/resources/2018/8/2018-nmc-horizon-report

Çağlayan, E. (2024). Alternative e-assessment in language classes: Recent experiences and future directions. *Journal of Qualitative Research in Education, 38,* 234-257. https://doi.org/10.14689/enad.38.1893

Cao, Y., Li, S., Liu, Y., Yan, Z., Dai, Y., Yu, P. S., & Sun, L. (2023). *A comprehensive survey of ai-generated content (AIGC): A history of generative AI from GAN to CHATGPT.* https://doi.org/10.48550/arXiv.2303.04226

Chen, X., Gao, Y., Tang, W., Guan, J., & Ryoo, J. (2024a). AI application in foreign language literature: ChatGPT's impact and skill enhancement. *European Journal of Contemporary Education and E-Learning, 2*(2), 3-18. https://doi.org/10.59324/ejceel.2024.2(2).01

Chen, X., Hu, Z., & Wang, C. (2024b). Empowering education development through AIGC: A systematic literature review. *Education and Information Technologies,* 1-53. https://doi.org/10.1007/s10639-024-12549-7

Chiu, T. K. (2024). Future research recommendations for transforming higher education with generative AI. *Computers and Education: Artificial Intelligence, 6,* 100197. https://doi.org/10.1016/j.caeai.2023.100197

Dashtestani, R. (2014). English as a foreign language—teachers' perspectives on implementing online instruction in the Iranian EFL context. *Research in Learning Technology,* 22, 1-15. https://doi.org/10.3402/rlt.v22.20142

Divekar, R. R., Drozdal, J., Chabot, S., Zhou, Y., Su, H., Chen, Y., Zhu, H., Hendler, J., & Braasch, J. (2022). Foreign language acquisition via artificial intelligence and extended reality: Design and evaluation. *Computer Assisted Language Learning, 35*(9), 2332-2360. https://doi.org/10.1080/09588221.2021.1879162

Diwan, C., Srinivasa, S., Suri, G., Agarwal, S., & Ram, P. (2023). AI-based learning content generation and learning pathway augmentation to increase learner engagement. *Computers and Education:Artificial Intelligence, 4,* 100110. https://doi.org/10.1016/j.caeai.2022.100110

Guo, D., Chen, H., Wu, R., & Wang, Y. (2023). AIGC challenges and opportunities related to public safety: A case study of ChatGPT. *Journal of Safety Science and Resilience, 4*(4), 329-339. https://doi.org/10.1016/j.jnlssr.2023.08.001

Haristiani, N. (2019). Artificial intelligence (AI) chatbot as language learning medium: An inquiry. *Journal of Physics. Conference Series, 1387*(1), 12020. https://doi.org/10.1088/1742-6596/1387/1/012020

Hu, B. (2010). The challenges of Chinese: A preliminary study of UK learners' perceptions of difficulty. *Language Learning Journal, 38*(1), 99-118. https://doi.org/10.1080/09571731003620721

Kasneci, E., Seßler, K., Küchemann, S., Bannert, M., Dementieva, D., Fischer, F., ... Kasneci, G. (2023). ChatGPT for good? On opportunities and challenges of large language models for education. *Learning and Individual Differences, 103,* 102274. https://doi.org/10.1016/j.lindif.2023.102274

Kirkpatrick, D. (1996). Revisiting Kirkpatricks four-level-model. *Training & Development, 1*, 54-57.

Kohnke, L., Moorhouse, B. L., & Zou, D. (2023). Exploring generative artificial intelligence preparedness among university language instructors: A case study. *Computers and Education: Artificial Intelligence, 5*, 100156. https://doi.org/10.1016/j.caeai.2023.100156

Krashen, S. D. (1982). *Principles and practice in second language acquisition.* Pergamon.

Lee, L. W., Dabirian, A., McCarthy, I. P., & Kietzmann, J. (2020). Making sense of text: Artificial intelligence-enabled content analysis. *European Journal of Marketing, 54*(3), 615-644. https://doi.org/10.1108/EJM-02-2019-0219

Li, B., & Siu, I. (2009). Developing students' affective attitude in primary English foreign language classrooms in the People's Republic of China. *Pedagogy, Culture & Society, 17*(2), 221-236. https://doi.org/10.1080/14681360902934467

Lim, W. M., Gunasekara, A., Pallant, J. L., Pallant, J. I., & Pechenkina, E. (2023). Generative AI and the future of education: Ragnarök or reformation? A paradoxical perspective from management educators. *The International Journal of Management Education, 21*(2), 100790. https://doi.org/10.1016/j.ijme.2023.100790

Liu, M., Ren, Y., Nyagoga, L. M., Stonier, F., Wu, Z., & Yu, L. (2023). Future of education in the era of generative artificial intelligence: Consensus among Chinese scholars on applications of ChatGPT in schools. *Future in Educational Research, 1*(1), 72-101. https://doi.org/10.1002/fer3.10

Lu, Y., Yu, J.L., & Chen, P.P. (2023). Educational application and prospect of generative artificial intelligence: Taking ChatGPT system as an example. *Chinese Journal of Distance Education, 43*(4), 24-31. doi:10.13541/j.cnki.chinade.20230301.001

Maja, M. M. (2023). Teachers' perceptions of integrating technology in rural primary schools to enhance the teaching of English first additional language. *Journal of Curriculum Studies Research, 5*(1), 95-112. https://doi.org/10.46303/jcsr.2023.8

Moorhouse, B. L., & Kohnke, L. (2024). The effects of generative AI on initial language teacher education: The perceptions of teacher educators. *System, 122*, 103290. https://doi.org/10.1016/j.system.2024.103290

Nazaretsky, T., Ariely, M., Cukurova, M., & Alexandron, G. (2022). Teachers' trust in AI - powered educational technology and a professional development program to improve it. *British Journal of Educational Technology, 53*(4), 914-931. https://doi.org/10.1111/bjet.13232

Seker, M., & Dincer, A. (2014). An insight to students' perceptions on teacher feedback in second language writing classes. *English Language Teaching, 7*(2), 73-83. http://dx.doi.org/10.5539/elt.v7n2p73

Sharadgah, T. A., & Sa'di, R. A. (2022). A systematic review of research on the use of artificial intelligence in English language teaching and learning (2015-2021): What are the current effects? *Journal of Information Technology Education: Research*, 21. http://dx.doi.org/10.28945/4999

Shen, H. H. (2018). Chinese as a second language reading: lexical access and text comprehension. In C. Ke (Ed.), *Routledge handbook of Chinese second language acquisition* (pp. 134-150). Routledge. https://doi.org/10.4324/9781315670706

Swain, M. (1985). Communicative competence: Some roles of comprehensible input and comprehensible output in its development. In S. Gass & C. Madden (Eds.), *Input in second language acquisition* (pp. 235-253). Newbury Press.

Tafazoli, D. (2024). Exploring the potential of generative AI in democratizing English language education. *Computers and Education: Artificial Intelligence,* 100275. https://doi.org/10.1016/j.caeai.2024.100275

U.S. Department of State. (n.d.). *Foreign language training.* https://www.state.gov/foreign-language-training/

Wang, C., Wang, Y., & Zou, B. (2024). Revolutionising EFL pedagogy: Innovative strategies for integrating GAI (ChatGPT) into language teaching. *Journal of Language Teaching, 4*(1), 1-7. https://doi.org/10.54475/jlt.2024.004

Wang, M., Wang, M., Xu, X., Yang, L., Cai, D., & Yin, M. (2023). Unleashing ChatGPT's power: A case study on optimizing information retrieval in flipped classrooms via prompt engineering. *IEEE Transactions on Learning Technologies, 17,* 629-641. https://doi.org/10.1109/TLT.2023.3324714

Wang, Y., Wu, J., Chen, F., Wang, Z., Li, J., & Wang, L. (2024). Empirical assessment of AI-powered tools for vocabulary acquisition in EFL instruction. *IEEE Access, 12,* 131892-131905. https://doi.org/10.1109/ACCESS.2024.3446657

Yang, S., Yang, S., & Tong, C. (2023). In-depth application of artificial intelligence-generated content AIGC large model in higher education. *Adult and Higher Education, 5*(19), 9-16. https://doi.org/10.23977/aduhe.2023.051902

Yang, T., & Gao, M. (2020). Studying Chinese characters in a web-based learning environment: A case study of Swedish university students. *Chinese Language Teaching Methodology and Technology, 3*(2). https://engagedscholarship.csuohio.edu/cltmt/vol3/iss2/2

Part II.
Research studies on emerging technologies in second language acquisition

Chapter 5

Impact of AI on language teaching pedagogy: a participatory action study

Xiaying Li
Fordham University

Su-Je Cho
Fordham University

Belle Li
Purdue University

Abstract

This participatory action research study explores integrating AI-powered learning tools in language classrooms, focusing on collaborative efforts with language teachers to design and implement AI-enhanced learning interventions. Employing a mixed-method approach, the study aims to understand how AI can potentially impact teachers' pedagogical practices and instructional strategies. Data collection includes quantitative measures through pre-test and post-test assessments and qualitative insights from teacher reflections, students' reflections, and observations. Central to this research is the collaborative engagement between learners, World Language teachers, and researchers in co-designing and implementing AI-driven learning interventions. The objectives are twofold: firstly, to co-create with educators in developing diverse, AI-integrated teaching strategies that foster greater classroom engagement and inclusivity, and secondly, to evaluate the effectiveness of these strategies in improving language learners' vocabulary learning outcomes. The results show that students improve their vocabulary acquisition and become more motivated, and the teacher improves her pedagogy and instruction making it more engaging.

Keywords: Participatory action research; Quizlet; GenAI; World language learning; Heritage learners

* * *

1. Introduction

Vocabulary is the foundation of the four language skills (listening, speaking, reading, and writing). Insufficient vocabulary can result in difficulties in expression, communication, and comprehension (Atalan & Subasi, 2023). Students may face challenges in written and oral tasks, which require productive skills due to inadequate lexical knowledge (Sofian & Salam, 2015; Yang, 2015). Scholars in the vocabulary field have explored various teaching techniques and pedagogical strategies to improve vocabulary instruction (Pourakbari & Biria, 2015; Schmitt, 2000; Zou, 2017). Although progress has been shown in vocabulary instruction, research continues to explore effective methods and approaches (Hulstijn & Laufer, 2001; Karalık, 2016). The need to reform vocabulary teaching in the formal language classroom is prominent (Marmol & Sánchez-Lafuente, 2013). Therefore, researchers, practitioners, and language teachers have focused on effective pedagogy and instructional strategies to enhance learners' vocabulary skills (Liu, 2009). Traditional vocabulary instructions, such as rote memorization and the chalkboard approach, are no longer the most effective way to teach vocabulary, although they still have some value. Instead, vocabulary instruction has shifted mainly to technology-driven instruction (Nejati et al., 2018).

Given the complexity of second language (L2) vocabulary acquisition, researchers have made significant efforts to improve it, for example through explicit vocabulary instruction (Tahir, 2017); learning vocabulary through extensive reading (Daskalovska; 2018 Sato, 2012); using word predictor to learn vocabulary in academic listening (Vidal, 2003); and using technology to learn vocabulary (Palmeira et al., 2020; Yin, 2003). Recently, technology-based vocabulary instruction has become a prominent trend in the language classroom (Gavranovic, 2019; Groot, 2000; Huszti & Barta, 2024). Additionally, integrating technology into the classrooms offers various benefits for learners in the 21st century (Ritzhaupt et al., 2012). Therefore, teachers should also modify their pedagogical instruction and enhance their teaching practices by learning and utilizing various AI-driven learning tools.

However, AI-driven pedagogy has not yet been thoroughly researched in vocabulary teaching because it requires teachers to equip themselves with strong digital and technology skills, and requires educators to be able to align AI with the curriculum. In this study, we employed participatory action research (PAR) to assist the participant, a Spanish teacher, in developing and reforming her pedagogy.

The PAR, rooted in the ideas of Kurt Lewin and Paulo Freire, provides a valuable means of exploring these challenges. PAR emphasizes collaboration between researchers and practitioners, bridging the gap between theoretical research and practical application in education (Kemmis et al., 2014). The unique contribution of this study lies in its use of PAR to position teachers as co-designers of AI-enhanced pedagogy. Rather than treating GenAI as a one-size-fits-all tool, this study explores how a high school Spanish teacher collaboratively reimagined her practices by incorporating Quizlet AI-powered features. This study had two main objectives: (1) To work closely with a high school Spanish teacher to design and implement AI-based learning pedagogy and strategies that improve classroom engagement and inclusivity; the focus was on using AI to tackle long-standing challenges in language teaching, and (2) To thoroughly evaluate the success of these collaboratively developed strategies in enhancing vocabulary learning outcomes. This process will employ a mixed-methods approach, utilizing both quantitative data to measure linguistic improvement and qualitative feedback to gain insight into student experiences and engagement.

By situating this research within the PAR framework, we aim to generate findings that are not only theoretically robust but also directly applicable to the everyday realities of language classrooms. This approach aligns with recent calls in the field for increased practitioner involvement in research on AI and education, recognizing that successful integration of these technologies requires the active participation and insights of educators (Moorhouse et al., 2024).

By foregrounding teacher-AI collaboration as a site of critical AI literacy, this chapter extends the disclosure on AI in education beyond the technical potential of instructional innovation. It demonstrates how educators can effectively integrate AI tools to achieve curriculum goals, alleviate workload, and transform classroom dynamics. In doing so, it offers a model for rethinking teacher roles and pedagogical strategies in the age of AI.

2. Literature review

2.1. AI in language education

The integration of AI in language education has prompted fundamental changes in pedagogical approaches. Research indicates that AI tools are reshaping traditional teaching methodologies, with inquiry-based learning emerging as a particularly effective framework for AI integration (Aslan et al., 2025; Yang et al., 2025). Yeh (2024) demonstrates how this approach creates more dynamic and student-centered learning environments, while simultaneously developing students' critical thinking skills. The transformation extends to assessment practices, where generative AI is creating opportunities for more authentic evaluation methods and reimagining how student learning is measured (Foung et al., 2024).

The implementation of AI in language education faces several significant challenges, particularly in the area of teacher preparation and professional development (Chokkappagari, 2025). Research by Kohnke et al. (2024) reveals that language instructors often experience significant technostress when implementing AI tools. In contrast, Moorhouse (2024) notes that beginning teachers frequently lack confidence in effectively using generative AI. Varying beliefs among EFL university teachers about integrating ChatGPT and other large language models compound these challenges (Gao et al., 2024).

2.2. AI in vocabulary learning

Vocabulary and grammar acquisition have also benefited from AI integration. Zhang and Huang (2024) observed that chatbots based on large language models effectively enhanced both receptive and productive vocabulary knowledge. Similarly, Alhammad (2024) reported improvements in students' understanding of literary themes, symbols, and character development, which indirectly contributed to language proficiency. Moreover, language learners can enlarge their vocabulary using an interactive storytelling chatbot (Alsadoon, 2021). Language learners can enhance their vocabulary input through AI functions, such as images, dictionaries, concordance, and first language (L1) translation tools. In this chatbot, language learners can enhance their vocabulary learning through images, dictionaries, concordance, and L1 translation tools provided by AI. Baker et al. (2020) researched the effectiveness of the intelligent tutoring systems (ITS) in learners' vocabulary improvement. They used a cluster randomized design, recruited Spanish-speaking learners as participants, and found that participants who used ITS scored significantly higher than those who did not use ITS on their word measurements.

2.3. Quizlet in vocabulary learning

Quizlet is an AI-powered online flashcard tool offering more than 18 languages (Quizlet, 2019). It not only offers free online material for students to study but also provides some features for teachers to conduct formative and summative assessments. Users can access the Quizlet tool on phones or computers through iOS or Android apps. They can choose from a variety of topics or create their own learning sets. Users can choose a variety of topics or create their own learning set. Although the program displays a prompt on one side and provides the answer on the other, similar to traditional flashcards, Quizlet enhances the learning experience by aligning the words with pictures and audio (Quizlet, 2019).

Quizlet can enhance learners' vocabulary learning by providing multisensory approaches. Rather than viewing Quizlet as a platform, many treat its affordances as task conditions that instantiate well-attested SLA mechanisms. Flashcards with audio and images serve as input enhancements and multimodal cues, supporting the noticing of form–meaning mappings (Tuite et al., 2012; Browne & Culligan, 2008). Matching and test modes require effortful retrieval, aligning with research on retrieval practice and the involvement load perspective for vocabulary consolidation (Azabdaftari & Mozaheb, 2012; Basoglu & Akdemir, 2010). Live/Checkpoint activities provide immediate feedback (confirmation, correction) and interactional pressure that promote uptake; team modes add peer scaffolding and collaborative dialogue around lexical choice and form. Q-Chat shifts practice to homework as guided output, extending time-on-task and enabling spaced practice across contexts.

2.4. AI and engagement

A consistent theme across studies is the positive impact of AI tools on learner engagement and motivation. Meyer et al. (2024) reported increased task motivation and positive emotions among students using AI-generated feedback. These results align with those of Ghafouri et al. (2024), who observed that a ChatGPT-based protocol enhanced both teachers' self-efficacy and learners' writing skills, thereby creating a more positive learning environment. The interactive and personalized nature of AI tools appears to be a key factor in this increased engagement. Fathi and Rahimi (2024) noted that the non-judgmental nature of AI-generated feedback reduced writing anxiety among learners, promoting a more comfortable learning experience.

2.5. AI and pedagogy

The integration of AI in language learning is reshaping pedagogical approaches and the roles of teachers. Moorhouse et al. (2024) highlighted the need to develop teachers' professional competence in generative AI, suggesting a shift towards teachers as facilitators of AI-human interactions. This imperative aligns with observations by Woo et al. (2024), who noted varying levels of prompt engineering abilities among students, implying a need for teacher guidance in effectively utilizing AI tools. Foung et al. (2024) emphasized the importance of a balanced approach in using various AI tools, integrating both traditional AI and generative AI in language learning contexts. This call for balance suggests an evolving role for teachers in curating and guiding AI-assisted learning experiences.

Research on teacher cognition reveals that beliefs, prior experience, and contextual constraints influence how teachers adopt new tools and redesign tasks. In AI-mediated contexts, shifts often include (a) moving from teacher-fronted explanation to facilitation of human–AI–learner interaction, (b) rebalancing planning time from materials creation toward diagnosis and differentiation informed by AI-generated data, and (c) rearticulating teacher identity and agency as designers who set pedagogic purposes and ethical guardrails. We therefore treat GenAI not as a replacement for expertise but as a catalyst for belief revision and practice reconfiguration, a stance we test empirically in this study.

2.6. Ethical considerations

Despite the promising results, several challenges and ethical considerations have emerged. Technical issues and the risk of over-reliance on AI are frequently cited concerns (Foung et al., 2024; Tsai et al., 2024). The need for practical training on AI use for both teachers and students was highlighted across multiple studies (Karataş et al., 2024; Moorhouse et al., 2024).

Ethical considerations, particularly those related to data privacy, have become increasingly important. Flowerdew (2024) and Tsai et al. (2024) raised questions about the balance between AI assistance and independent learning, emphasizing the need for guidelines on the ethical use of AI in educational settings.

While the current landscape is promising, researchers have identified several gaps in the literature. Several studies have called for more longitudinal research to understand the long-term effects of AI integration in language learning (Mahapatra, 2024; Meyer et al., 2024). There is also a need for larger sample sizes and more diverse contexts to enhance the generalizability of findings (Polakova & Klimova, 2024; Yavuz et al., 2024). Additionally, as AI tools continue to evolve, there is a growing need for research on developing AI literacy among both educators and learners. This need includes understanding how to effectively craft prompts, critically evaluate AI-generated content, and integrate AI tools into broader language learning strategies.

We position this study at the intersection of SLA, teacher cognition, and critical digital pedagogy. From SLA, we draw on: (a) noticing: learners benefit when input is made salient and attention is directed to lexical form–meaning mappings; (b) interaction & feedback: uptake emerges through opportunities for response, verification, and correction; (c) scaffolding & mediation: teacher/peer and tool supports enable performance within the learner's zone of proximal development; and (d) retrieval and spaced practice: repeated, effortful recall consolidates vocabulary. Teacher cognition frames how beliefs and knowledge shape classroom decisions and change over time; we examine how GenAI-supported tasks reconfigure the teacher's roles, planning moves, and assessment practices. Finally, critical digital pedagogy orients design toward equity, agency, and workload justice, asking whether the use of AI expands participation (e.g., among heritage learners), reduces unnecessary labor, and keeps pedagogy, not tools, at the center. These lenses guide task design (Section 4), operationalize expected mechanisms (Table 5.1), and anchor our analysis and discussion of learning and teaching outcomes.

Task/Feature	SLA construct activated	Pedagogical rationale	What we measured/observed
Flashcards (audio + image) as Do-Now	Noticing; input enhancement; multimodality	Make lexical form–meaning salient; build initial mapping efficiently	Faster recognition in class; higher post-test item accuracy for taught sets
Match/Test modes with retries	Retrieval practice; spaced repetition	Effortful recall consolidates vocabulary	Pre→post vocabulary gains; reduced errors on repeated items
Quizlet Live (solo → team)	Feedback; interaction; scaffolding	Immediate correctness signals; peer coaching; collaborative dialogue	More hand-raising, on-task behavior; student comments about "helping teammates"
Checkpoint (timed quiz)	Focused attention; output under time pressure	Elicit quick production/selection; diagnostic data for CF	Teacher reports of targeted re-teaching; formative alignment with goals
Q-Chat for homework	Guided output; extended practice	Move practice beyond class; maintain spacing	Student reflections on independent review; homework completion rates

Table 5.1. SLA-informed design and pedagogical outcomes of Quizlet-based vocabulary tasks

2.7. Participatory action research in education

PAR functions as a methodology for investigating and understanding communities with the explicit aim of enacting positive changes within them (Bergold & Thomas, 2012; Morales, 2019; Reason & Bradbury, 2008). PAR has evolved from its initial focus on social organization to a more comprehensive approach that emphasizes empowerment, critical reflection, and societal change (Morales, 2019). It has expanded its application to various fields, including education, community development, and social reform, consistently emphasizing the empowerment of participants and societal change. PAR can help bridge this gap by involving educators and learners in the research process, ensuring that theoretical developments are grounded in practical realities. Practically, PAR can help identify and address actual challenges faced by language learners and educators when integrating AI tools.

We adhered to a cyclical PAR process. PAR emphasizes a repetitive cycle of planning, action, observation, reflection, and revision, engaging participants actively to address real-world issues effectively (Kemmis et al., 2014; Whyte et al., 1989). This process goes beyond simply "doing" and reflecting afterward; it ensures that individuals make informed decisions before acting and promptly implement every significant insight (Morales, 2019).

3. Research Questions

1. How, if at all, does integrating Quizlet AI-powered tools help world language teachers with their pedagogical practice?
2. Does the use of AI-powered learning tools improve the vocabulary scores of students in Spanish classrooms?
3. How, if at all, does AI-integration instruction change classroom dynamics?

4. Methods

4.1. Study design

The researchers framed this study as a PAR project, which fundamentally relies on collaboration among researchers, teachers, and learners. Together, we co-designed and implemented the use of AI-powered learning tools in the language classroom. Through this collaborative approach, we aimed to ensure that the interventions are contextually relevant and aligned with the educational needs and goals of the participants.

4.2. Mixed-method approach

The researchers used a mixed-method approach to address the research questions. We collected qualitative data through in-depth interviews with world language teachers and learners, supplemented by a systematic collection of reflective notes from teachers and detailed observation notes taken during classroom sessions. These interviews explored the integration of AI tools in pedagogical practices, revealing stakeholders' motivations, challenges, and perceived outcomes. Teachers' reflections added insights into the daily impact and adjustments made in response to the integration of AI tools. At the same time, observation notes provided real-time data on classroom dynamics, student engagement, and interactions with AI technologies.

Quantitatively, we assessed the impact of AI-powered learning interventions using pre-tests and post-tests, with a focus on vocabulary acquisition. These tests measured changes in students' academic performance.

4.3. Participants

We conducted this collaborative action research with Susan (pseudonym), a Spanish teacher, in a charter school in the Bronx, New York. Originally from Peru, Susan has been teaching at this school for two years and has a total of 25 years of experience teaching Spanish. She instructs students from 9th grade to 12th grade.

The demographics of her students are mainly Latino students and or African American students from low-income families. Half of these students are heritage learners, meaning they are second- or third-generation Americans who can understand some Spanish due to parental communication but are not fluent in the language. The rest of the students are African Americans whose first language and or home language is not Spanish. The group comprises 53 students ranging in age from 14 to 16 years old.

4.4. Instrumentation

After collaborative discussions with teachers and learners, we adopted Quizlet as an AI-powered tool to implement in the classroom.

Quizlet is a widely recognized learning platform that provides a diverse range of interactive study materials. It enables users to tailor their learning experiences and access an extensive database of user-generated content, facilitating efficient and effective learning across multiple subjects and disciplines. In this study, Quizlet was utilized as a central tool in a Spanish language classroom to promote vocabulary learning and enhance student engagement, as detailed in Table 5.1 and Figure 5.1. The platform was employed daily, where students interacted with digital flashcards to familiarize themselves with new vocabulary. To assess student learning, the teacher utilized Quizlet Live and the Checkpoint feature, both of which provide timely feedback.

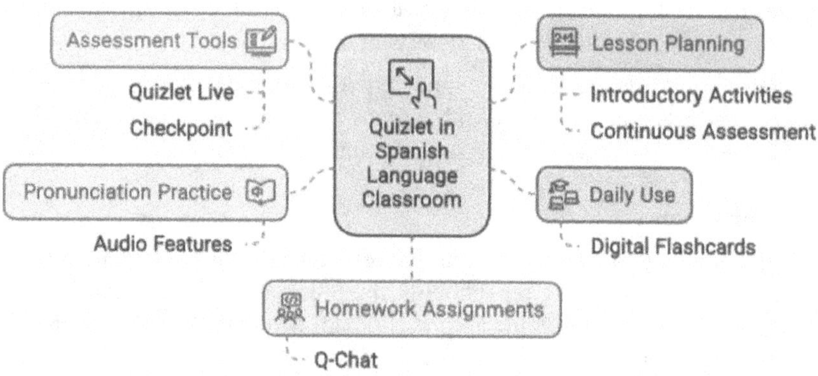

Figure 5.1. Integration of Quizlet tools in the Spanish language classroom

Additionally, Quizlet's audio capabilities were used extensively for pronunciation practice. The use of Quizlet Q-Chat was specifically designated for homework assignments, allowing students to engage in post-class review at home. The integration of Quizlet into the classroom extended to lesson planning, where it was used not only as an introductory activity but also as a continuous assessment tool throughout the lesson.

4.5. Procedures

Our study's procedures demonstrate a rigorous application of the PAR process. The process commenced with the identification of core issues faced by both educators and learners in a Bronx charter school's Spanish language program. Initial conversations with the Spanish teacher revealed several challenges, including (1) insufficient support from the school system for language instruction, (2) difficulties in engaging students in vocabulary learning and review in the classroom, and (3) a lack of adequate practice and support for students outside of school hours.

With all of these obstacles in world language classrooms, the research team, along with the Spanish teacher and a student representative, decided to use an AI-powered tool in the classroom as the solution. At the outset, the researchers discussed several possible approaches and began the intervention using Kahoot as the primary tool. We started with Kahoot as the intervention tool. However, the Spanish teacher noted after two weeks that Kahoot usually took a long time to complete the whole vocabulary review because Kahoot shows one question at a time (a time-intensive single-question format), which is not an efficient way to review and reinforce vocabulary memory.

Then, we decided to adopt another AI-powered learning tool, Quizlet. Quizlet Live has no time frame for each question. Instead, students can answer the questions at their own pace, and the first students who answer the question 100% will be the champions. Quizlet also provides a Kahoot-similar function called checkpoint, which provides a time frame for each question. Moreover, Quizlet enables students to listen to the pronunciation of flashcards to review vocabulary and use the Q-chat, an AI function of Quizlet, to self-assess their progress.

After a two-week trial, feedback from the Spanish teacher and the student representative was overwhelmingly positive, leading to the decision to continue using Quizlet as a learning tool in the Spanish classroom. The Spanish teacher used Quizlet at least once per day, depending on the lesson planning. For example, if this is an introductory class, the Spanish teacher incorporated the Quizlet by showing students the Quizlet flashcards and listening to the pronunciation. The Spanish teacher also checked for understanding by utilizing the Quizlet checkpoint and Quizlet Live functions, which provide students with immediate feedback and teachers with a snapshot of student performance.

The total time to adopt Quizlet in the Spanish classroom was four weeks; the researchers conducted two observations (one in the second week and one in the fourth week) in total to monitor the integration process and gather data, which included both the teacher's and students' reflections recorded in audio and written formats. The Spanish teacher regularly debriefed and reflected on pedagogical practices with researchers each week through in-person meetings with audio recordings. Finally, students were required to write a reflection at the end of the four weeks.

4.6. Measurement

To evaluate the effectiveness of Quizlet, we used archived data alongside pre-tests and post-tests designed by the Spanish teacher. These tests assessed vocabulary knowledge before and after the Quizlet integration, each containing 100 vocabulary items, with each item scored on a point basis for a total of 100 points. The researchers specifically designed this test for the project to assess participants' vocabulary learning progress. They selected vocabulary items from the *IB Spanish A* textbook.

4.7. Data analysis

In this study, we employed a comprehensive data analysis strategy to assess the impact of integrating an AI-powered learning tool (Quizlet) on both pedagogical practices and student outcomes in a language learning context.

We conducted a thematic analysis of the interview data from the Spanish teacher using NVivo to examine changes in pedagogy. We triangulated this with the teacher's reflective recordings to enhance the credibility and trustworthiness of the data. Quantitatively, we aimed to assess the impact of Quizlet on students' academic performance, specifically in vocabulary acquisition. To this end, we conducted a paired *t*-test on the students' pre-test and post-test scores. Additionally, to obtain a holistic view of the tool's impact, we triangulated these quantitative results with observation notes taken during classroom sessions.

5. Results

5.1. Impact on Spanish teachers' pedagogy

Two key pieces of data clarified how AI-powered learning tools impact the pedagogy of Spanish teachers. In the reflection, the Spanish teacher wrote:

"This is my first time using Quizlet in the classroom as an instructional tool. It completely changed my teaching methods. I used to be a very traditional teacher who only liked to use paper and pen only. I know my students sometimes do not really enjoy it, and they do not really remember the vocabulary. However, I see my students change after I use Quizlet in my classroom. The kids are very engaging now and always ask me if we are going to play Quizlet. This also makes me rethink teaching and learning. Students will always learn better when they are motivated and curious. Traditional ways cannot excite them anymore. Quizlet can be used with different functions, such as a rewards system, competition, collaboration, and AI Q&A functions. On the other hand, Quizlet also reduces my preparation time because of Quizlet."

In the interview data, we found that (1) with Quizlet-integrated instruction, the teacher shifted her pedagogy from teacher-centered to student-centered with enhanced use of technology; (2) Quizlet changed her lesson planning style; and (3) formative assessments aligned with the learning goal following the adoption of Quizlet. We employed NVivo thematic methods to identify patterns in the data (Table 5.2).

Themes	Sample Quotes
Shift to student-centered learning	"My teaching philosophy and practice have greatly changed after integrating Quizlet as part of my teaching. I reduce the direct instruction time while I have more independent practice and learning time for students." "I started a lesson with a Quizlet matching game, which allows my students to explore the words independently."
Technology-enhanced instruction	"I adapt my teaching styles from a traditional way to a more technology-driven way."

	"I started using less pen and paper. Instead, we tried to use more AI-powered learning tools to help students get immediate feedback."
Lesson planning change	"I am taking Quizlet as part of my lesson planning consideration. I use Quizlet as a do-now, a mid-point check for understanding, formal vocabulary assessments, etc." "Quizlet also reduced my lesson planning time. So, I can spend more time differentiating instruction rather than creating the resources from scratch." "Quizlet makes my lesson more targeted and relevant."
Formative assessment	"I can make a sound judgment of students' levels based on the Quizlet formative data, which gives me a more comprehensive overview of each students' performance." "My formative assessment is 100% aligned with my learning goals because the Quizlet generates the quiz of what we have learned from me."

Table 5.2. Interview data results

5.2. Students' academic performance

We conducted a paired *t*-test to examine students' Spanish vocabulary acquisition before and after integrating the AI-powered learning tool into the Spanish classroom.

Specifically, we explored whether pre-test scores differed from post-test scores. It is important to note that participants were not randomly sampled, thereby violating one of the assumptions of t-tests. The results suggested that there was a statistically significant difference in scores ($p < .001$) between pre-test ($M = 67.08$, $SD = 17.042$) and post-test ($M = 78.45$, $SD = 12.886$) scores. In other words, students improved from an average score of 67.08 to 78.45. Moreover, the effect sizes indicate that these differences were significant (Cohen's $d = 1.317$). Ultimately, these results suggest that students' vocabulary scores increase dramatically over time.

5.3. Classroom dynamic change

Based on the researchers' observation notes and students' reflections. We also noticed some changes in the classroom. In the observation notes, we wrote down students' engagement in the school. To be more specific, we tracked students' hand-raising, task completion, group discussions, and other relevant behaviors. We had two observations before using Quizlet as an instructional tool in the classroom, and two observations while using Quizlet as the instructional tool. We found that the Quizlet implementation in the classroom engages more students in the class. The observation notes appear in the appendix.

In addition, students' reflections demonstrate the change in classroom climate, showing that the innovative instruction in the Spanish class largely switched their learning motivation and joy (Table 5.3).

Student Reflections
"I usually don't like my Spanish class because it was so boring. But I feel like I like my Spanish class now that my Spanish teacher has used Quizlet in class. I look forward to doing Quizlet every day."
"It was so fun! I hope we will continue to use Quizlet to learn Spanish."
"Ms. Susan changed her teaching this week. I think this new teaching works very well for me. I don't need to use a pen and pencil to learn the vocabulary now. Ms. Susan makes this class so interesting, and I can remember more vocabulary."
"Quizlet is the best. We used Quizlet in the class to learn, remember, and review the vocabulary. We can also use Quizlet at home to review and practice the vocabulary. I love Quizlet flashcards and Q-chat. I always used the audio of flashcards to remind myself of the pronunciation. And I can listen to it multiple times until I get it. Also, I can use Q-chat to practice vocabulary with me and self-assess."
"I really like the Quizlet part. I think I am more engaged than before in the class. I raised my hands more and participated in the activity more. Quizlet is fun and also helps me learn."

Table 5.3. Student reflections

6. Discussion

The findings of this study extend beyond vocabulary enlargement to reveal broad pedagogical takeaways. The observed vocabulary gains and engagement patterns are consistent with tasks that increased noticing (Schmidt, 1990; Sharwood Smith, 1993) through salient, multimodal input (Mayer, 2009), retrieval via daily recall with feedback (Karpicke & Blunt, 2011; Roediger & Karpicke, 2006; see also Barcroft, 2007, for L2 vocabulary), and scaffolded interaction (Long, 1996; Swain, 1985; Vygotsky, 1978; Wood et al., 1976). Concomitantly, the teacher's reflections indicate a shift in cognition from "explainer" to designer–facilitator, aligning assessment and practice through frequent, low-stakes checks and targeted support (Borg, 2003, 2006). In PAR cycles (Kemmis & McTaggart, 2000; Reason & Bradbury, 2001), these redesigned routines consolidated new beliefs about efficacy; in our GenAI-mediated implementation, similar patterns surfaced in L2 vocabulary instruction (Wu & Pan, 2024), suggesting that GenAI can mediate sustainable changes in planning, assessing, and differentiating when decisions are theory-led rather than tool-led; central among these, AI literacy in cultivating reflective teaching practice.

The Spanish teacher's pedagogy underwent a significant shift due to the implementation of Quizlet. She transitioned from traditional teaching to technology-driven and student-centered teaching. Previously, the Spanish teacher used only pen and paper in class for students to learn Spanish. She has now incorporated Quizlet into her teaching and students' learning. Switching to a more up-to-date pedagogy, she balanced traditional teaching with technology-driven instruction. This new pedagogy and instruction not only help the teacher with her lesson planning to be more structured and effective but also engage students in her class. Through observation and students' reflection, her classroom is more inclusive, engaging, and welcoming because she provides different approaches to learning. Students are more excited to come to class compared to before, and the number of hand-raising incidents increases, as does the time spent on-task. Besides, her lesson planning time shrank after using Quizlet as one of the resources in the class, which eases her workload. For teacher educators, this case provides practical strategies for adaptation. First, begin with low-stakes AI tools, such as flashcards and vocabulary games, before moving to more complex AI functions. Second, provide a reflection prompt for teachers and students to critically evaluate the effectiveness of the AI tool critically, ensuring that AI adoption is not technocentric but pedagogically driven.

By critically engaging with Quizlet's functions, the Spanish teacher exercised agency in shaping her practice, rather than passively adopting the new technology. The results reinforce the idea that AI literacy is vital for teacher preparation and professional development. Teachers who learn to interrogate, critique, adapt, and repurpose AI tools can not only enhance students' learning outcomes but also reduce their workload and reclaim time to differentiate instruction. These skills contribute to the model of teacher agency in which AI supports rather than supplants. Moreover, the AI-integrated assignment in this study demonstrates the precise alignment with L2 teaching competencies by enhancing communicative skills, formative assessment practices, and lesson design.

Surprisingly, the Quizlet-integrated lessons also changed the classroom dynamic. After using Quizlet in the Spanish class, students' engagement and involvement increased. Students are more motivated and involved in the class. Students who tended to be inactive in the classroom before participated well in the class. Several participants directly pointed out that they love the new approach to learning, which makes them feel fun and motivated.

This study, however, has some limitations. First, this is not a large-scale study. The sample size was only 53, which was relatively small. Besides, this study was conducted only in a Spanish classroom. It would be more convincing to have this study in different world language classrooms.

Moreover, we learned that world language teachers lack support in pedagogy and instruction because every language is different, and professional development leaders do not have adequate knowledge of all the world languages. Thus, world language teachers employ the pedagogy and instruction with which they are familiar. Additionally, the world language department is often overlooked by the school administration team because it is not considered a core subject in K-12 education. On the other hand, the professional development offered by schools and school districts does not cover the topic of technology-driven pedagogy, which causes teachers to be unsure of how to incorporate advanced technology into their classrooms and adopt an updated teaching philosophy to engage students.

Schools and school districts should allocate more resources to the World Language Department and provide solution-oriented and innovative professional development for World Language teachers. In addition, teachers should be open-minded about advanced technology and adopt appropriate technology into classrooms to engage students, improve students' performance, and reduce teachers' workload.

Future studies should explore the transferability of this model to other world languages, grade levels, and economic and social contexts. Comparing multiple AI platforms may also provide insights into which features best support inclusivity and workload reduction.

In conclusion, this study underscores the potential of AI tools for rethinking classroom practice. By centering collaboration, reflection, and agency, AI-integrated pedagogy offers a pathway to be more engaging, inclusive, and innovative language learning.

References

Alhammad, A. I. (2024). The Impact of ChatGPT in Developing Saudi EFL learners' literature appreciation. *World Journal of English Language, 14*(2), 331-338. https://doi.org/10.5430/wjel.v14n2p331

Alsadoon, R. (2021). Chatting with AI bot: Vocabulary learning assistant for Saudi EFL learners. *English Language Teaching, 14*(6), 135. https://doi.org/10.5539/elt.v14n6p135

Aslan, S., Alyuz, N., Li, B., Durham, L. M., Shi, M., Sharma, S., & Nachman, L. (2025). An early investigation of collaborative problem solving in conversational AI-mediated learning environments. *Computers and Education: Artificial Intelligence, 100393*, https://doi.org/10.1016/j.caeai.2025.100393

Atalan, E., & Subasi, G. (2023). The use of Quizlet in teaching vocabulary to 9th grade EFL students. *Turkish Online Journal of Educational Technology (TOJET), 22*(3), 131–152. https://tojet.net/articles/v22i3/22312.pdf

Azabdaftari, B., & Mozaheb, M. A. (2012). Comparing vocabulary learning of EFL learners by using two different strategies: Mobile learning vs. flashcards. *The EUROCALL Review, 20*(2), 47–59. https://doi.org/10.4995/eurocall.2012.11377

Baker, D. L., Ma, H., Polanco, P., Conry, J. M., Kamata, A., Al Otaiba, S., Ward, W., & Cole, R. (2020). Development and promise of a vocabulary intelligent tutoring system for second-grade Latinx English learners. *Journal of Research on Technology in Education, 53*(2), 223–247. https://doi.org/10.1080/15391523.2020.1762519

Barcroft, J. (2007). Effects of opportunities for word retrieval during second language vocabulary learning. *Language Learning, 57*(1), 35–56. https://doi.org/10.1111/j.1467-9922.2007.00398.x

Basoglu, E. B., & Akdemir, O. (2010). A comparison of undergraduate students' English vocabulary learning: Using mobile phones and flashcards. *Turkish Online Journal of Educational Technology (TOJET), 9*(3), 1–7.

Bergold, J., & Thomas, S. (2012). Participatory research methods: A methodological approach in motion. *Historical Social Research/Historische Sozialforschung,* 37(4),191–222.

Borg, S. (2003). Teacher cognition in language teaching: A review of research on what language teachers think, know, believe, and do. *Language Teaching, 36*(2), 81–109. https://doi.org/10.1017/S0261444803001903

Borg, S. (2006). *Teacher cognition and language education: Research and practice.* Continuum.

Brown, C., & Culligan, B. (2008). Combining technology and IRT testing to build student knowledge of high-frequency vocabulary, *Japan Association Language Learning - Computer Assisted Language (JALT CALL) Journal, 43*(2), 2016.

Chokkappagari, R. (2025). Teacher development programs in language education: AI-driven linguistic perspectives on effective practices. *Infact: International Journal of Computers* 9(1), 1-12. https://doi.org/10.61179/infact.v9i01.647

Daskalovska, N. (2018). Extensive reading and vocabulary acquisition. In R. J. Ponniah & S. Venkatesan (Eds.), *The idea and practice of reading* (pp. 25-40). Springer. https://doi.org/10.1007/978-981-10-8572-7_3

Fathi, J., & Rahimi, M. (2024). Utilising artificial intelligence-enhanced writing mediation to develop academic writing skills in EFL Learners: A qualitative study. *Computer-Assisted Language Learning,* 1-40. https://doi.org/10.1080/09588221.2024.2374772

Flowerdew, J. (2024). Data-driven learning: From Collins COBUILD dictionary to ChatGPT. *Language Teaching, 58*(3), 1–18. https://doi.org/10.1017/S0261444824000144

Foung, D., Lin, L., & Chen, J. (2024). Reinventing assessments with ChatGPT and other online tools: Opportunities for GenAI-empowered assessment practices. *Computers and Education: Artificial Intelligence, 6,* 100250. https://doi.org/10.1016/j.caeai.2024.100250

Gao, Y., Wang, Q., & Wang, X. (2024). Exploring EFL university teachers' beliefs in integrating ChatGPT and other large language models in language education: A study in China. *Asia Pacific Journal of Education, 44*(1), 29-44. https://doi.org/10.1080/02188791.2024.2305173

Gavranović, V. (2019). Second language vocabulary Acquisition enhanced by the use of technology. Sinteza 2019: International Conference on information Technology and Data Related Research, Belgrade, Serbia. https://doi.org/10.15308/Sinteza-2019-621-626

Ghafouri, M., Hassaskhah, J., & Mahdavi-Zafarghandi, A. (2024). From virtual assistant to writing mentor: Exploring the impact of a ChatGPT-based writing instruction protocol on EFL teachers' self-efficacy and learners' writing skill. *Language Teaching Research,* 1–23. https://doi.org/10.1177/13621688241239764

Groot, J. M. (2000). Computer-assisted second language vocabulary acquisition. *Language Learning & Technology, 4*(1), 60–81. https://doi.org/10.64152/10125/25087

Hulstijn, J. H., & Laufer, B. (2001). Some empirical evidence for the involvement load hypothesis in vocabulary acquisition. *Language Learning, 51*(3), 539–558. http://dx.doi.org/10.1111/0023-8333.00164

Huszti, I. I., & Barta, K. A. (2024). Methods of teaching foreign language vocabulary online. *Academic Notes. Series: Pedagogical Sciences,* (9), 66-72. https://doi.org/10.59694/ped_sciences.2024.09.066

Karalık, T., & Merç, A. (2016). The effect of task-induced involvement load on incidental vocabulary acquisition. *Mustafa Kemal University Journal of Graduate School of Social Service, 13*(35), 77-92.

Karataş, F., Abedi, F. Y., Gunyel, F. O., Karadeniz, D., & Kuzgun, Y. (2024). Incorporating AI in foreign language education: An investigation into ChatGPT's effect on foreign language learners. *Education and Information Technologies, 29,* 19343-19366. https://doi.org/10.1007/s10639-024-12574-6

Karpicke, J. D., & Blunt, J. R. (2011). Retrieval practice produces more learning than elaborative studying with concept mapping. *Science, 331*(6018), 772–775. https://doi.org/10.1126/science.1199327

Kemmis, S., & McTaggart, R. (2000). Participatory action research. In N. K. Denzin & Y. S. Lincoln (Eds.), *Handbook of qualitative research* (2nd ed., pp. 567–605). Sage.

Kemmis, S., McTaggart, R., & Nixon, R. (2014). *The action research planner: Doing critical participatory action research.* Springer. https://doi.org/10.1007/978-981-4560-67-2_5

Kohnke, L., Zou, D., & Moorhouse, B. L. (2024). Technostress and English language teaching in the age of generative AI. *Educational Technology & Society, 27*(2), 306–320. https://doi.org/10.30191/ETS.202404_27(2).TP02

Liu, J. H. (2009). The integration of CALL to vocabulary teaching and learning. *US-China Foreign Language, 7*(7), 60-64. https://doi.org/10.17265/1539-8080/2009.07.010

Long, M. H. (1996). The role of the linguistic environment in second language acquisition. In W. C. Ritchie & T. K. Bhatia (Eds.), *Handbook of second language acquisition* (pp. 413–468). Academic Press. https://doi.org/10.1016/B978-012589042-7/50015-3

Mahapatra, S. (2024). Impact of ChatGPT on ESL students' academic writing skills: A mixed methods intervention study. *Smart Learning Environments, 11*(9). https://doi.org/10.1186/s40561-024-00295-9

Mármol, G. A., & Sánchez-Lafuente, Á. A. (2013). The involvement load hypothesis: The effect on vocabulary learning in primary education. *Revista Española de Lingüística Aplicada, 26*, 11–24.

Mayer, R. E. (2009). *Multimedia learning* (2nd ed.). Cambridge University Press. https://doi.org/10.1017/CBO9780511811678

Meyer, J., Jansen, T., Schiller, R., Liebenow, L. W., Steinbach, M., Horbach, A., & Fleckenstein, J. (2024). Using LLMs to bring evidence-based feedback into the classroom: AI-generated feedback increases secondary students' text revision, motivation, and positive emotions. *Computers and Education: Artificial Intelligence, 6*, 100199. https://doi.org/10.1016/j.caeai.2023.100199

Moorhouse, B. L. (2024). Beginning and first-year language teachers' readiness for the generative AI age. *Computers and Education: Artificial Intelligence, 6*, 100201. https://doi.org/10.1016/j.caeai.2024.100201

Moorhouse, B. L., Wan, Y., Wu, C., Kohnke, L., Ho, T. Y., & Kwong, T. (2024). Developing language teachers' professional generative AI competence: An intervention study in an initial language teacher education course. *System, 125*, 103399. https://doi.org/10.1016/j.system.2024.103399

Morales, M. P. (2019). Participatory action research (PAR) in education. In C. A. Mertler (Ed.), *The Wiley handbook of action research in education* (pp. 317–341). Blackwell.

Nejati, E., Jahangiri, A., & Salehi, M. R. (2018). The effect of using computer-assisted language learning (CALL) on Iranian EFL learners' vocabulary learning: A Quizlet study. *Cypriot Journal of Educational Sciences, 13*(2), 351–362.

Palmeira, E. G., Martin, V. B., Gonçalves, V. B., Moraes, I. D., Lamounier Júnior, E. A., & Cardoso, A. (2020). The use of immersive virtual reality for vocabulary acquisition: A systematic literature review. In *Anais do XXXI Simpósio Brasileiro de Informática na Educação* (pp. 532-541). https://doi.org/10.5753/cbie.sbie.2020.532

Polakova, P., & Klimova, B. (2024). Implementation of AI-driven technology into education: A pilot study on the use of chatbots in foreign language learning. *Cogent Education, 11*(1), 2355385. https://doi.org/10.1080/2331186X.2024.2355385

Pourakbari, A. A., & Biria, R. (2015). Efficacy of task-induced involvement in incidental lexical development of Iranian senior EFL students. *English Language Teaching, 8*(5), 122–131. https://doi.org/10.5539/elt.v8n5p122

Quizlet. (2019). *Our mission.* Retrieved October 26th, 2025, from https://quizlet.com/mission

Reason, P., & Bradbury, H. (2001). *Handbook of action research: Participative inquiry and practice.* Sage.

Reason, P., & Bradbury, H. (2008). *Sage handbook of action research: Participative inquiry and practice* (2nd ed.). Sage. https://doi.org/10.4135/9781848607934

Ritzhaupt, A. D., Dawson, K., & Cavanaugh, C. (2012). An investigation of factors influencing student use of technology in K-12 classrooms using path analysis. *Journal of Educational Computing Research, 46*(3), 229–254. https://doi.org/10.2190/EC.46.3.b

Roediger, H. L., III, & Karpicke, J. D. (2006). Test-enhanced learning: Taking memory tests improves long-term retention. *Psychological Science, 17*(3), 249–255. https://doi.org/10.1111/j.1467-9280.2006.01693.x

Sato, F. (2012). Vocabulary acquisition through extensive reading. *Accents Asia, 6*(1), 52-69.

Schmidt, R. (1990). The role of consciousness in second language learning. *Applied Linguistics, 11*(2), 129–158. https://doi.org/10.1093/applin/11.2.129

Schmitt, N. (2000). *Vocabulary in language teaching.* Cambridge University Press.

Sharwood Smith, M. (1993). Input enhancement in instructed SLA: Theoretical bases. *Studies in Second Language Acquisition, 15*(2), 165–179. https://doi.org/10.1017/S0272263100011943

Sofian, H., & Salam, U. (2015). Improving students' writing ability in narrative text by using picture series in SMA. *Jurnal Pendidikan dan Pembelajaran, 4*(4), 1–12. https://doi.org/10.26418/jppk.v4i4.9893

Swain, M. (1985). Communicative competence: Some roles of comprehensible input and comprehensible output in its development. In S. Gass & C. Madden (Eds.), *Input in second language acquisition* (pp. 235–253). Newbury House.

Tahir, M. H. (2017). Research and trends in the studies of explicit vocabulary instructions and vocabulary acquisition. *Journal of Educational Studies, 3(2).*

Tsai, C.-Y., Lin, Y.-T., & Brown, I. K. (2024). Impacts of ChatGPT-assisted writing for EFL English majors: Feasibility and challenges. *Education and Information Technologies* 29, 22427–22445. https://doi.org/10.1007/s10639-024-12722-y

Tuite, K., Sercu, L., Peters, E. (2012). Multimodal input and vocabulary learning: The role of images and sound in form-meaning mapping. Language Learning & Technology, 16(3), 145-163.

Vidal, K. (2003). Academic listening: A source of vocabulary acquisition? *Applied Linguistics, 24* (1):56-89. https://doi.org/10.1093/applin/24.1.56

Vygotsky, L. S. (1978). *Mind in society: The development of higher psychological processes.* Harvard University Press.

Whyte, W., Greenwood, D., & Lazes, P. (1989). Participatory action research. *American Behavioral Scientist, 32,* 513–551. https://doi.org/10.1177/0002764289032005003

Woo, D. J., Guo, K., & Susanto, H. (2024). Exploring EFL students' prompt engineering in human-ai story writing: An activity theory perspective. *Interactive Learning Environments,* 1–20. https://doi.org/10.1080/10494820.2024.2361381

Wood, D., Bruner, J. S., & Ross, G. (1976). The role of tutoring in problem solving. *Journal of Child Psychology and Psychiatry, 17*(2), 89–100. https://doi.org/10.1111/j.1469-7610.1976.tb00381.x

Yang, Y. I. (2015). An investigation of Chinese junior high school teachers' and students' attitudes towards EFL Writing. *International Journal of Research Studies in Education, 5*(2), 59-76. https://doi.org/10.5861/ijrse.2015.1209

Yang, M., Jiang, S., Li, B., Herman, K., Luo, T., Moots, S. C., & Lovett, N. (2025). Analysing nontraditional students' ChatGPT interaction, engagement, self-efficacy and performance: A mixed-methods approach. *British Journal of Educational Technology, 56*(5), 1973-2000. https://doi.org/10.1111/bjet.13588

Yavuz, F., Çelik, Ö., & Yavaş Çelik, G. (2024). Utilizing large language models for EFL Essay grading: An examination of reliability and validity in rubric-based assessments. *British Journal of Educational Technology, 56*(1), 150-166. https://doi.org/10.1111/bjet.13494

Yeh, H. C. (2024). The synergy of generative AI and inquiry-based learning: Transforming the landscape of English teaching and learning. *Interactive Learning Environments, 33*(1), 88-102. https://doi.org/10.1080/10494820.2024.2335491

Yin, Z. (2003). On vocabulary acquisition. *Journal of Xinxiang Teachers College.* https://en.cnki.com.cn/Article_en/CJFDTOTAL-XXSF200306038.htm

Zhang, Z., & Huang, X. (2024). The impact of chatbots based on large language models on second language vocabulary acquisition. *Heliyon, 10,* e25370. https://doi.org/10.1016/j.heliyon.2024.e25370

Zou, D. (2017). Vocabulary acquisition through cloze exercises, sentence-writing and composition-writing: Extending the evaluation component of the involvement load hypothesis. *Language Teaching Research, 21*(1), 54–75. https://doi.org/10.1177/1362168816652418

Appendix: classroom observation notes

Date	Quizlet Integration	Observation Notes	Engagement Tracker
5/8/2024	No	1. This classroom visit was announced. We communicated with the Spanish teacher before and scheduled the date and time that she preferred. We stayed in the classroom for 20 minutes. The Spanish teacher was teaching the Unit on "Community." 2. In these 20 minutes, the Spanish teacher taught vocabulary through a reading text. The reading text was in the textbook. 3. The Spanish teacher asked some questions about the correct answer for multiple-choice reading.	Three students raised their hands within twenty minutes to answer questions.
5/18/2024	No	1. This classroom observation was communicated. However, we did not tell the teacher the exact time we would observe her class. 2. This class focused on vocabulary review. The teacher used some physical vocabulary flashcards to review vocabulary with students. 3. Students who sat at the back of the classroom cannot see the flashcards. They did not seem very engaged in the activity. 4. Students should raise their hands if they know how to say the vocabulary based on the teachers' flashcards (only pictures). 5. We observed the class for around 20 minutes.	Ten students in total who sat in the front (first row and second row) raised their hands within twenty minutes.
5/28/2024	Yes	1. This observation was scheduled with the Spanish teacher at a specific time and date.	100% of students focused on the Quizlet game during the solo

		2. The Spanish teacher used Quizlet as a Do-now to review the vocabulary. Before she taught the class, 100% of students used their Chromebooks to join Quizlet Live. 3. The Spanish teacher had two rounds of Quizlet Live. The first round was a solo game. The second round was a team collaboration game.	round. Students were actively using their notes to help them select the correct answers. Students said "Yes" when they got the correct answer when the Quizlet provided them immediate feedback. No students opt out during the team collaboration game. Students in each group help their teammates to find the correct answers.
6/4/2024	Yes	1. This observation was announced, but the Spanish teacher was not told the specific time of observation. 2. The Spanish teacher was using Quizlet to teach new vocabulary. She first used flashcard functions to teach vocabulary by playing the audio of the word from Quizlet and projecting the Quizlet flashcard on the whiteboard. 3. Then, the Spanish teacher used matching games to check for understanding.	Thirty students were raising their hands when the matching game was happening in five minutes. 100% of students repeated after the audio, even students who sat at the back of the classroom.

Chapter 6

AI chatbots in EFL teaching and learning: insights from teachers and learners

Nebahat Badem

Kahramanmaras Sutcu Imam University

Omer Ozer

Adana Alparslan Turkes Science and Technology University

Abstract

This chapter offers a comprehensive analysis of the perspectives held by Turkish EFL teachers and learners concerning the integration of AI chatbots into language learning and teaching. The study involved five teachers from the School of Foreign Languages at a state university in Turkey, who taught English courses, and five students from two different language levels within the same institution. Semi-structured interviews were conducted to elicit detailed insights from participants about their experiences and perspectives on using AI chatbots for foreign language teaching and learning. The findings indicate that ChatGPT is the most frequently used AI tool among both teachers and students, primarily employed to support the writing and speaking skills of EFL learners. Additionally, participants noted that AI chatbots enable them to complete tasks in significantly less time than would normally be required. A commonly expressed concern among participants from both groups is the risk of over-reliance on AI chatbots for language proficiency development. The chapter concludes by identifying areas for further research and potential improvements related to AI chatbots in foreign language education.

Keywords: AI chatbots; Language teaching; Language learning; Teacher perspectives; Student perspectives

* * *

1. Introduction

Artificial intelligence (AI) has made significant advancements, particularly since 2022, as the ways people access information have undergone rapid evolution. Students, likewise, have rapidly integrated this technology into academic practices. AI offers a wide range of opportunities for students, teachers, and researchers alike. In particular, AI chatbots, the focus of this study, are currently making the process of accessing and acquiring knowledge even easier for these educational stakeholders. Foreign language education is an area where chatbots are utilized for various purposes, particularly in language skills development, as they can interact with humans through voice, text, or a combination of both.

The incorporation of chatbots into language education brings both opportunities and challenges. Key advantages of AI chatbots for language learners include creating educational content, reducing stress and anxiety, providing personalized practice (Adamopoulou & Moussiades, 2020; Zhang & Huang, 2024), and offering accessibility anywhere and at any time (Begum et al., 2024; Duong & Suppasetseree, 2024; Haristiani, 2019). Additionally, AI chatbots facilitate repetition (Haristiani, 2019), adapt to individual learning pace (Yuen & Schlote, 2024), and can improve language proficiency and communicative competence (Shikun et al., 2024; Yuen & Schlote, 2024; Zhang et al., 2023). For language teachers, these tools facilitate exam preparation and evaluation, reduce feedback time (Mohamed, 2024), and assist with administrative tasks (Korda et al., 2024).

Nonetheless, AI chatbots present notable limitations. Potential drawbacks include overreliance on these bots (Mohamed, 2024; Ozer, 2024; Patty, 2024), a novelty effect and loss of motivation (Fryer et al., 2017), frustration with inaccurate voice recognition (Shikun et al., 2024), and privacy as well as academic ethical concerns (Chan, 2023; Mohamed, 2024). Furthermore, in regions where educational inequality persists, the gap between students in these areas and those in developed countries is thought to widen even further. Fitria (2021) identified the relatively high costs of AI as a problem for language learners in developing countries, particularly when they aim to access premium AI functionalities.

2. Literature review

As interest in the study of AI chatbots in education and especially in the context of foreign language education has increased substantially in recent academic literature, an expanding body of research has investigated this topic across varied educational settings and methodological approaches. In this section, we highlight studies that examine AI tools from a language education perspective, and we follow with a discussion of several systematic reviews on the topic.

In language education, some studies focused exclusively on either the advantages or disadvantages of chatbots through descriptive research designs (Alrajhi, 2024; Mohamed, 2024; Yuen & Schlote, 2024). In one such study, employing a mixed-methods approach with 143 English as a foreign language (EFL) learners of varying language proficiency levels, Alrajhi (2024) investigated these learners' experiences and perspectives on a text-based pedagogical chatbot. Overall, the study yielded positive results; however, there were no significant results in terms of language proficiency. Nevertheless, students believed that a certain level of language proficiency was necessary to interact with the chatbot. Additionally, regarding the future use of the chatbot, students had mixed dispositions about continuing to use the chatbot. This indicated a need to balance the novelty of such tools with attentive task design was suggested to maintain learners' motivation and interest (Alrajhi, 2024). Similarly, in another study, which qualitatively investigated the perceptions of 10 EFL faculty members (Mohamed, 2024), varying opinions regarding the advantages, concerns, the position of ChatGPT in the future, and the implementation of ChatGPT in classes as an addition to traditional practices were elicited. Mohamed (2024) argues that traditional language teaching methods are still essential for a personal, human-centered approach to language learning that improves conversational skills and understanding of culture. Overall, Mohamed revealed "real-time interaction and feedback, personalized language instruction, a vast knowledge base, natural language processing capabilities, the ability to generate human-like responses and engagement," and "cost-effectiveness" as the advantages while revealing "a lack of human connection and personalization, the potential for inaccurate or inappropriate responses, difficulties in understanding context and meaning, limitations in pronunciation and intonation feedback, cultural insensitivity, cheating tool," and "technical limitations" (p. 3205) as the disadvantages. Additionally, Yuen and Schlote (2024) investigated native English speakers' experiences of using AI to learn other languages through mobile apps and AI. Based on survey results from Canadian post-secondary students and conducted follow-up interviews with 20 of them, they concluded that apps are perceived as intriguing and convenient for L2 learning and the integration of AI for conversation-based simulations or speech recognition would lead to more adaptive and personalized second language learning (L2; Yuen & Schlote, 2024). Taken together, all these studies describe the advantages and disadvantages of chatbots in language education, reflecting general perceptions among the participants involved.

With a shift in focus, studies such as those by Fryer et al. (2017, 2019) highlighted the comparison of chatbots to human teachers or human partners in interaction. Fryer et al. (2017) compared chatbots with humans in speaking activities in a language course over 12 weeks in Japan with an experimental design. Ultimately, they concluded that interest in interactions with chatbots declined, suggesting two possible explanations: the novelty effect and a lack of authenticity. Building on their previous work, Fryer et al. (2019) added a qualitative element through open-ended discussion questions. They concluded with codes for merits and demerits for both chatbot and human partners, summarizing they both had their benefits such as convenience, technical benefits, communication ease or social benefits, and drawbacks such as communication problem, technical difficulty or social problems (Fryer et al., 2019). These two studies revealed both strengths and weaknesses of chatbots and human teachers or partners over the other without specifically promoting one.

Instead of describing or evaluating the relationship between the existing chatbots and language learners, some studies moved beyond and developed context-specific chatbots themselves and evaluated these chatbots from various perspectives (Dhivvya & Karnati, 2024; Han et al., 2023). For instance, Han et al. (2023) developed a tool called RECIPE, a ChatGPT-integrated writing tool designed to support student essay revision. of the study involved 213 students and seven instructors from South Korea. They concluded with overall positive results from both students and instructors with some suggestions for the platform (Han et al., 2023). Similarly, Dhivvya and Karnati (2024) developed BuddyBot, which they later tested with a questionnaire. Similar to the results presented by Han et al., they also elicited overall positive results, and the participants seemed likely to use the bot in the future in their English language learning process. The study concluded with a favorable user perception, foregrounding the role of AI-powered chatbots in language learning (Dhivvya & Karnati, 2024). Both studies present significant contributions in terms of highlighting the customization features of chatbots for different pedagogical needs in different educational contexts.

Besides these studies with a wider focus on language learning, some research has targeted specific areas such as vocabulary acquisition using AI chatbots (Zhang & Huang, 2024) or the development of speaking skills (Duong & Suppasetseree, 2024; Shikun et al., 2024; Zhang & Huang, 2024). They conducted a mixed-methods quasi-experimental study to examine the impact of chatbots on second-language vocabulary acquisition. The study focused on receptive and productive vocabulary learning through the use of chatbots over an eight-week period with 52 intermediate-level high school students whose first language is Mandarin. Their pre-test - post-test, and delayed post-test showed that the experimental group significantly outperformed the control group in receptive, productive, and incidental vocabulary acquisition. Similarly, Shikun et al. (2024) investigated students' speaking skill improvement using AI chatbots among low- and intermediate-level middle school students in China and employed a pre-test and post-test design, complemented by a qualitative component. The results were overall positive, showing improvement between pre- and post-tests, although there were some drawbacks, such as disappointment with AI chatbots' misunderstandings and failures in speech recognition (Shikun et al., 2024). Finally, Duong and Suppasetseree (2024) examined the effects of an AI voice chatbot (Andy) on the English-speaking skills of A2-level Vietnamese students. They conducted a quasi-experimental mixed-methods study with a pre-test and post-test design over an 8-week implementation period. Using questionnaires and interviews, they concluded that students showed improvements in speaking skills, accuracy, and fluency. The students' attitudes towards using AI voice chatbot were mostly positive, though there were some noted limitations, similar to those reported by Shikun et al. Duong and Suppasetseree's (2024) findings, like those of Zhang and Huang (2024), also underscore incidental vocabulary learning. These studies refer to the specific features of chatbots that can be used for distinctive areas of language learning, if need be, rather than a general approach.

In the Turkish setting, we have identified only one experimental study. Punar Özçelik and Yangın Ekşi (2024) investigated the impact of ChatGPT on writing development, testing register learning over a four-week period with 11 Turkish EFL students. They collected observational data from one of the researchers and conducted interviews with the students, finding that the participants deemed ChatGPT useful for learning formal register, though unnecessary for informal register. Additionally, their study did not elicit conclusive results concerning neutral register (Özçelik & Ekşi, 2024). This study, however, was limited in scope, testing a single skill with a small number of participants over a short period.

In this rapidly changing area of research, there are also some systematic reviews about the history, usage, benefits, and challenges of AI chatbots and the use of these chatbots in education in particular (Adamopoulou & Moussiades, 2020; Bozkurt & Sharma, 2024; Haristiani, 2019; Gökçearslan et al., 2024; Jinming & Daniel, 2024; Korda et al., 2024). Some of these reviews identified AI chatbots that may support language learners and teachers in instructional settings. In their systematic literature review, Gökçearslan et al. (2024) analyzed the prevalence of widely used AI chatbots and reported that ChatGPT ranked first, followed by Ellie. In addition, Duong and Suppasetseree (2024) cite Kim et al. (2019), who identify a range of AI voice chatbots used for

English language learning, including "ELIZA, ALICE, Cleverbot, Elbot, Talk to Eve, Replika, Lyra, Andy English Bot, Poker Friend, and Mondly" (p. 296). Moreover, Jinming and Daniel (2024) argue that chatbots such as Replica, Ellie, Andy, Cleverbot, and ChatGPT4 are distinctive due to their superior natural language processing abilities, which, in return, promote a more natural interaction. Finally, Korda et al. (2024) include IBM Watson Assistant, ChatGPT, Scribe Sense, Duolingo, Coursera, and Woebot as AI chatbots in education. Providing such lists of recommended chatbots may support language teachers in making informed pedagogical decisions and enhancing their instructional practices.

3. The present study

The introduction and the subsequent incorporation of AI technology in its most recent form into language education has been so rapid particularly in the last few years (Berry, 2023) – especially with the launch of ChatGPT on November 30, 2022 – that it has been difficult to adapt to the process and make necessary changes in teaching curricula or material development instantly (Barrot, 2023; Ozer, 2024). Therefore, a considerable number of teachers and students have nevertheless begun to use these chatbots for educational purposes. While they benefit from AI chatbot technology with overall positive effects, this usage has also raised some issues that are potentially detrimental to teaching and learning. Given the recent introduction of this technology, it is crucial at this stage to capture the dual perspectives of both language learners and teachers by closely examining their relatively early experiences.

Considering that AI will likely play an increasingly active role in the future of language education, more research is needed in this realm. Our study makes a timely contribution to the literature for two primary reasons. First, it examined the experiences and perspectives of both university students and teachers. Second, the study deepens understanding of participants' perspectives and experiences through a qualitative approach, uncovering specific uses of AI that either hinder or facilitate learning. With this aim in mind, we designed our study in guidance of the following research questions: (1) What are EFL teachers' experiences and perspectives on the use of AI chatbots as part of their language teaching and (2) What are EFL students' experiences and perspectives on the use of AI chatbots as part of their language learning?

4. Methodology

4.1 A Qualitative Exploration of Teachers' and Students' Experiences with AI Chatbots

Our study employed a qualitative research design to understand and describe the lived experiences and perspectives of EFL students and teachers. By focusing on participants' detailed perspectives regarding their use of AI chatbots, we conducted interviews, which Cohen et al. (2002) suggest are potent tools for obtaining comprehensive answers, including those addressing complex and nuanced issues, while allowing for spontaneity. To gain insights into participants' experiences and perceptions, we applied a critical lens to understand how AI is transforming language education and to identify potential risks it poses for students and learners.

4.2. Participants

Our study included ten participants: five EFL teachers with a mean age of 35.4 years and five EFL students with a mean age of 20.4 years, all from the School of Foreign Languages at a state university in southern Turkey. All five teachers reported a mean of 12.2 years of teaching experience and were actively teaching English courses at the time of data collection. Similarly, all five students had been attending classes for eight months and were about to complete their studies in the program. Four of the student participants were at the level of B1+ and one at C1 according to the Common European Framework of Reference (Council of Europe, 2001).

Using convenience sampling, the available teachers and students were individually contacted and interviewed during data collection. This sampling strategy was chosen to ensure accessibility and timely data collection from participants with relevant teaching and learning experience, all of whom had prior experience with AI chatbots.

4.3. Instrumentation

We conducted semi-structured interviews with participating EFL teachers and students through live individual videoconferences, resulting in a total of 10 interviews. We developed eight open-ended questions to guide our investigation (Table 6.1). For the students, the same questions were reworded and simplified to elicit their perspectives. In crafting the interview questions, the researchers aimed to align them with the methodologies and findings of previous studies. Specifically, questions one and two were adapted from items in Chan's (2023) quantitative survey. Questions six and seven, which address the challenges and benefits that students might experience, were adapted and extensively modified from questions previously developed by Ozer (2024). The remaining questions were informed by findings from previous research on human-computer interaction, including studies by Mohamed (2024) and Fryer et al. (2019), which highlight that AI chatbots may serve as valuable tools for complementing and enhancing foreign language teaching practices. Research by Chen et al. (2023) provided rationale, suggesting that listening and speaking skills can be significantly impacted by AI usage.

Guiding Interview Questions
Which artificial intelligence chatbots do you use most?
Which knowledge or skills of your students do you prefer AI chatbots to support?
How do you use these robots? Can you describe it with an example?
Which skills do you think your students develop the most while interacting with AI robots?
What is your stance on using voice and/or text AI robots to improve your students' language skills?
What are the difficulties your students encounter when using robots? How are you trying to overcome these challenges?
What is the biggest benefit you have experienced when using AI chatbots to contribute to your students' learning?
Do you ever think that AI chatbots are not useful for your students' language development?

Table 6.1. Guiding interview questions

The interviews with the participants were held in Turkish—L1 for all participants—so that they would not have problems with the language barrier while expressing their ideas and experiences. The purpose was to eliminate any potential limitations, especially for the student participants.

4.4. Data collection and analysis

After participants gave their consent, interviews were scheduled and conducted. Data were collected at the end of the 2023–2024 academic year, between May 16, 2024, and June 17, 2024, through video interviews. The interviews with teachers ranged from 13 to 28 minutes in length, while the interviews with students lasted between 8 and 15 minutes.

After data collection, we transcribed the participants' interview responses and analyzed the transcripts using a qualitative content analysis approach. Initially, we independently reviewed the data to become familiar with it, thoroughly rereading and studying it. Subsequently, we individually identified initial codes. We applied an inductive coding strategy to allow themes to emerge directly from the data, thereby avoiding preexisting categories. We then convened remotely to discuss and refine the emerging codes, ensuring consistency and addressing any discrepancies. The researchers developed themes collaboratively by grouping related codes and examining patterns across participant responses to ensure accurate representation of shared experiences and perspectives. Finally, we interpreted the meaning of these agreed-upon themes in relation to our research questions.

5. Findings

Regarding the AI chatbots used by teachers, we discovered that teachers predominantly utilize ChatGPT (4 participants) and Gemini (2 participants). A male teacher with 10 years of teaching experience remarked, "I usually use ChatGPT. I've also tried Gemini, but I don't use it very often." Similarly, students primarily rely on ChatGPT and GliGlish, as exemplified by one female student from the Translation and Interpreting department who revealed, "I use ChatGPT most frequently" (Table 6.2).

Teachers	*n*	Students	*n*
ChatGPT	4	ChatGPT	4
Gemini	2	GliGlish	1

Table 6.2. AI chatbots employed by participants for educational purposes.

Data from the interviews revealed that each participant group employs AI chatbots differently to enhance student learning. Teachers identified two main strategies: creating engaging learning activities and facilitating brainstorming sessions on stimulating topics (Table 6.3). One male teacher with ten years of teaching experience illustrated his approach to designing engaging learning activities by stating, "As I mentioned earlier, I primarily use artificial intelligence to enhance the materials I've prepared to improve my students' speaking and writing skills." Similarly, a female teacher with 14 years of teaching experience highlighted the role of AI chatbots in facilitating brainstorming, noting, "I used to use this to generate ideas [...] for my students in reading and writing classes." A male teacher with 10 years of teaching experience highlighted the potential use of AI to generate model essays and reading passages, stating, "I'm telling ChatGPT, 'Hey, can you write me a 100-word piece about this that has these specific qualities?'" Similarly, another teacher with five years of experience emphasized how AI can simplify the feedback process, noting, "The AI provides grammar feedback, allowing you to focus on offering comments on the content itself."

Teachers	*n*
Developing engaging learning activities	4
Facilitating the brainstorming of stimulating topics	4
Creating model essays for writing courses	3
Simplifying the process of delivering feedback	3

Table 6.3. Teachers' AI utilization strategies to enhance student learning

For students, creating model essays for writing courses, searching for information on a topic, and error correction in their sentences were the top three strategies (Table 6.4). Using AI to generate model paragraphs and essays in writing classes was a common strategy among students, as exemplified by one female student from the Computer Engineering department who stated, "I ask the AI about a topic first, then I ask it to write sample essays on that topic, and I read those." This student further explained that AI also enables students to search for and find specific information on the web. She added, "Sometimes there are English abbreviations or other specific information. I ask ChatGPT for the meanings of those, and it gathers some information." A male student from the Artificial Intelligence Engineering department highlighted its benefit in helping him find his mistakes: "Sometimes, I use AI to show the grammatical errors in my writing." Another male student from the Software Engineering department also mentioned that AI helps him practice speaking in English. He said, "When I prepare for my classes beforehand, I ask ChatGPT, 'What do you think about this topic?' and it responds to me in English. Or I start by asking, 'Would you like to discuss this topic with me?'"

Students	*n*
Creating model essays for writing courses	3
To gather and synthesize information on a topic	3
Error correction	3
Speaking practice	1

Table 6.4. Students' utilization strategies to enhance student learning.

Interestingly, regarding the language skills and components perceived as most benefiting from AI chatbot usage, both teachers and students shared almost identical views. Both groups believed that writing was the most advantageous skill, followed by speaking, vocabulary, grammar, listening, reading, and pronunciation. (Table 6.5). Notably, the student group did not mention listening. A male teacher with 10 years of teaching experience observed, "I think students could use it to improve their writing skills. They could get suggestions about something they've written. They could upload something they've written to ChatGPT and ask it to show them any grammar mistakes." Similarly, a male student from the Artificial Intelligence Engineering department remarked, "I mostly use AI to improve my writing and reading." Additionally, a male student from the Computer Engineering department remarked, "When learning a foreign language, the area where we students struggle the most is speaking. We find it very difficult to speak. I primarily use AI for one-on-one speaking practice and to help me prepare assignments."

Teachers	*n*	**Students**	*n*
Writing	4	Writing	5
Speaking	3	Speaking	2
Vocabulary	3	Vocabulary	2
Grammar	3	Grammar	1
Reading	1	Reading	1
Pronunciation	1	Pronunciation	1
Listening	1		

Table 6.5. Language skills and components perceived as benefiting most from AI chatbot usage

Regarding how participants perceive the benefits of AI usage, teachers and students hold differing opinions. The responses from three students highlighted a single concept: the opportunity to receive personalized guidance for improving language production. They also emphasized the instantaneous nature of their interactions with AI. One male student from the Software Engineering department remarked, "The biggest benefit for me is in terms of practice. I can't always practice with my friends. But when you're learning a new language, constant practice is necessary. I think using artificial intelligence is really beneficial in this regard." He further noted, "You can't always be with friends you get along with all the time, but thanks to this, you can practice the language at any time, even before bed."

Teachers' responses, on the other hand, reveal that the most frequently cited benefit they perceive is the time efficiency gained in preparing lessons and developing learning materials (Table 6.6). As one male teacher with ten years of teaching experience exemplified, "Sometimes, there's not enough time to prepare something new, and that's when ChatGPT can be really helpful." A female teacher with 12 years of experience further highlighted this advantage, stating, "I used to create my own texts to teach vocabulary [to my students] and spent a lot of time on it. [...] Thanks to artificial intelligence, I can now prepare them in a much shorter time." Another area where teachers benefit from AI is in simplifying the process of generating feedback for their students. A female teacher with 14 years of teaching experience remarked: "I used artificial intelligence to get [suggestions] on the feedback I created for my students in my writing class." Similarly, teachers emphasize that students themselves can receive instant language-use suggestions. A female teacher with five years of teaching experience stated, "This applies to writing classes as well. Providing detailed feedback to every student is challenging. However, students can write about additional topics and promptly review the grammar feedback section. This allows them to identify their mistakes." The same teacher emphasized that AI chatbots can promote more inclusive language education, particularly for students from lower socioeconomic backgrounds. She noted, "Not every student has the means to afford paid programs consistently. In this regard, I believe artificial intelligence offers a more equitable opportunity for students."

Teachers	*n*
Time efficiency in preparing lessons and developing materials	3
Simplifying the process of delivering feedback	2
Personalized instant guidance for improving student language production	2
Advancing equal opportunities for students	1

Table 6.6. Teachers' perceived benefits from AI usage

The responses from both groups highlight some differences in perceived AI-related challenges (Table 6.7). Teachers identified the two most pressing concerns as a potential rise in academic dishonesty and insufficient language development due to over-reliance on AI chatbots. Regarding academic integrity concerns, one participant with 20 years of teaching experience remarked, "If something is directly generated by artificial intelligence, it's not something you produced but something it produced. In fact, there's no difference between this and asking a friend to do your homework." Another teacher, with ten years of teaching experience, expressed concern that a student might become overly reliant on AI, which could negatively impact language development. He stated, "If a student doesn't clarify their own goals, doesn't know their weaknesses and strengths, and doesn't know what they want to do, using AI is like spoon-feeding. Whether I wrote it for the student or ChatGPT wrote it, it's the same thing." Three teachers also noted that inadequate student autonomy hinders effective AI usage. A male teacher with ten years of teaching experience, for example, stated, "I believe that, since our students are generally not very autonomous, they may rely more on artificial intelligence to achieve their goals without putting in much effort." Additionally, two teachers pointed out that students often struggle with crafting effective AI prompts. Emphasizing the importance of accurate prompt formulation, one female teacher with five years of experience stated, "The prompt they give [to the AI] needs to be accurate so that they can get a correct result."

Teachers	*n*
Increase in academic dishonesty cases	4
Deficient language development due to overreliance on AI chatbots	4
Insufficient student autonomy	3
Inability to craft effective AI prompts	2

Table 6.7. Teachers' perceived AI-related challenges

The students mentioned similar perceived challenges to those of teachers, but they also emphasized that AI could provide them with inaccurate or misleading information (Table 6.8). Consequently, they should exercise even greater caution when verifying the information. For instance, one female student from the Translation and Interpreting department said, "I have doubt about whether I can trust the information provided by artificial intelligence. I need to do more careful research based on the information it gives." To exemplify how the use of AI might be detrimental to language development, one female student from the Computer Engineering department remarked, "Information acquired through effort is difficult to forget, but information that is easily accessible, like that provided by AI, may be easily forgotten. When I research something myself, I consult multiple sources, which makes the information more memorable. However, when I use AI, it doesn't work that way." Another common challenge was students' difficulty in obtaining the intended output from the AI, as it often failed to align with their expectations. For example, one male student from the Computer Engineering department noted, "I'm trying to write prompts [for the AI] in English. Sometimes, I don't know how to express what I want to say in English. [...] I can't seem to convey what I want."

Students	n
Inaccurate or misleading information	2
Deficient language development due to overreliance on AI chatbots	2
Inability to craft effective AI prompts	2

Table 6.8. Students' perceived AI-related challenges

Both teachers and students expressed expectations regarding the integration of AI into foreign language learning. Three teachers highlighted the importance of AI literacy education for both students and educators. For instance, a female teacher with five years of experience suggested, "We should consider implementing AI literacy programs for both teachers and students." One female student from the Computer Engineering department shared her concern about understanding her teacher's stance on AI usage in coursework before deciding whether to use it. She explained: "One teacher was against us using ChatGPT, while another encouraged it, saying, 'Why aren't you using artificial intelligence? It could be very beneficial for your language development.' [...] Whether we can or cannot use AI, and to what extent, we decide based on our teachers' guidance."

6. Discussion

Our study set out to explore the experiences and perspectives of EFL teachers and students regarding the use of AI chatbots as part of their language teaching and learning. As presented in the previous section, the findings were predominantly positive, though teachers exhibited a cautious approach to students' use of AI chatbots, which aligns with and builds on existing literature (Mohamed, 2024; Patty, 2024).

Firstly, the most used AI chatbot by the participants in this study was ChatGPT, adding to the literature with similar results from the Turkish context. This result was also reported by Gökçearslan et al. (2024) in their systematic literature review. As argued by Jinming and Daniel (2024), the reason why the participants in this study also resorted to ChatGPT over other chatbots could be attributed to its superior natural language processing ability in addition to its user-friendliness and widespread usage.

The data also provided us with many AI utilization strategies by the participants. The most frequent AI utilization strategies by the teachers were developing engaging learning activities and facilitating the brainstorming of stimulating topics which were similarly reported by Adamopoulou and Moussiades (2020). In addition, our participants also mentioned simplifying the process of delivering feedback as a strategy for their AI utilization. Similarly, Mohamed (2024) reported that helping decrease feedback time was one of the advantages of using AI in education. From the students' perspectives, gathering and synthesizing information on a topic resonated with Mohamed (2024), who described ChatGPT as a vast knowledge base. Also, our students used AI chatbots to create model essays for writing which was also reported by Alrajhi (2024) and Han et al. (2023), and to practice their speaking similarly to Shikun et al. (2024) and Duong and Suppasetseree (2024). The reason why AI chatbots were initially perceived as a tool to improve writing skills than speaking skills could be due to their technological nature. In addition, students often view language learning as a process heavily focused on reading and writing before fully developing their speaking skills. This perception may explain their emphasis on writing more over speaking. Additionally, the voice chatbots could necessitate subscription plans which is why participants are mostly inclined toward text chatbots. To sum up, these similar results on the usage strategies of AI chatbots regardless of the different contexts of these studies lead to a common point in their use in language learning and teaching, expanding the common knowledge base.

Language skills that are mostly targeted with the use of AI chatbots were found to be writing and speaking in this study similar to those by Alrajhi (2024), Mohamed (2024), Han et al. (2023), Shikun et al. (2024), and Duong and Suppasetseree (2024). The next two language components that are mentioned by both teachers and students are vocabulary and grammar. Similar to our study, Zhang and Huang (2024) also reported overall positive effects of AI chatbots on vocabulary learning. It is noteworthy that students did not mention listening skill at all while they mentioned reading and pronunciation as the last two language skills they improved through the use of AI chatbots, similar to their teachers. The improvement in speaking and fluency skills was also found by Duong and Suppasetseree although some drawbacks were also reported by their participants. The reliance on reading and writing more than listening and speaking is once again highlighted here, supporting existing literature.

Participants' perceived benefits from AI usage in this study are consistent with their AI utilization strategies. Benefits reported in this study include AI chatbots' time efficiency in material development, simplified feedback delivery, personalized guidance, and the provision of equal opportunities. These results are complimentary to the existing research as they were also elicited by Ozer (2024), Mohamed (2024), and Yuen and Schlote (2024). In addition, Fryer et al. (2019) highlighted chatbots as free and online, which could offer opportunities for language learners globally.

In terms of the challenges with AI chatbot usage in language teaching and learning, our data indicated that an increase in academic dishonesty cases emerged as the primary concern from the teachers' perspectives. Mohamed (2024) similarly identified this as one of the disadvantages of the AI-based chatbot, ChatGPT. Both teachers and students in this study reported deficiencies in language development stemming from an overreliance on AI chatbots. Overreliance on these bots could be a serious issue, especially for students as it would lead them to laziness not only in their language learning but also overall academic studies, a concern echoed previously by Mohamed, Ozer (2024), and Patty (2024). Insufficient student autonomy was one of the concerns reported by teachers in this study, which was also mentioned by Ozer as a lack of awareness about the usage. The teachers emphasized the importance of students being autonomous learners to make the most of AI chatbots and their features in their language learning. Finally, both teachers and students in this study acknowledged their difficulty in crafting appropriate prompts for AI chatbots. This could be partly because of the proficiency levels of the students as mentioned by themselves both in this study and Alrajhi (2024). The students in that study believed a certain level of language proficiency was necessary to interact with the chatbot in the first place. In their systematic review, Jinming and Daniel (2024) similarly reported findings from Chen et al. (2023) that the language proficiency of the learners is likely to influence the topics that they can discuss with chatbots. It is noteworthy to present complementary results from the Turkish context alongside the other distinct contexts in which the aforementioned studies were conducted.

Finally, both teachers and students in this study expressed some expectations for AI to enhance their foreign language teaching and learning. Teacher participants mentioned a need for AI literacy training for both students and teachers while student participants emphasized a need for explicit school wide AI policy. Similar challenges, such as a lack of awareness regarding the usage and a lack of institutional support, were also identified by Ozer (2024). As stated earlier, a significant number of students and teachers worldwide have begun to use AI chatbots in their educational lives without adequate guidance or a solid foundation. Addressing the expectations of foreign language learners and teachers, from their immediate environments to broader contexts, could help mitigate some of these issues.

Our study set out to explore the experiences and perspectives of EFL teachers and students regarding the use of AI chatbots as part of their language teaching and learning. As presented in the previous section, the findings were predominantly positive, though teachers exhibited a cautious approach to students' use of AI chatbots, which aligns with and builds on existing literature (Mohamed, 2024; Patty, 2024).

Firstly, the most used AI chatbot by the participants in this study was ChatGPT, adding to the literature with similar results from the Turkish context. This result was also reported by Gökçearslan et al. (2024) in their systematic literature review. As argued by Jinming and Daniel (2024), the reason why the participants in this study also resorted to ChatGPT over other chatbots could be attributed to its superior natural language processing ability in addition to its user-friendliness and widespread usage.

The collected data provided us with many AI utilization strategies by the participants. The most frequent AI utilization strategies by the teachers were developing engaging learning activities and facilitating the brainstorming of stimulating topics which were similarly reported by Adamopoulou and Moussiades (2020). In addition, our participants also mentioned simplifying the process of delivering feedback as a strategy for their AI utilization. Similarly, Mohamed (2024) reported that helping decrease feedback time was one of the advantages of using AI in education. From the students' perspectives, gathering and synthesizing information on a topic resonated with Mohamed (2024), who described ChatGPT as a vast knowledge base. Also, our students used AI chatbots to create model essays for writing which was also reported by Alrajhi (2024) and Han et al. (2023), and to practice their speaking similarly to Shikun et al. (2024) and Duong and Suppasetseree (2024). The reason why AI chatbots were initially perceived as a tool to improve writing skills than speaking skills could be due to their technological nature. In addition, students often view language learning as a process heavily focused on reading and writing before fully developing their speaking skills. This perception may explain their emphasis on writing more over speaking. Additionally, the voice chatbots could necessitate subscription plans which is why participants are mostly inclined toward text chatbots. To sum up, these similar results on the usage strategies of AI chatbots regardless of the different contexts of these studies lead to a common point in their use in language learning and teaching, expanding the common knowledge base.

Language skills that are mostly targeted with the use of AI chatbots were found to be writing and speaking in this study similar to those by Alrajhi (2024), Mohamed (2024), Han et al. (2023), Shikun et al. (2024), and Duong and Suppasetseree (2024). The next two language components that are mentioned by both teachers and students are vocabulary and grammar. Similar to our study, Zhang and Huang (2024) also reported overall positive effects of AI chatbots on vocabulary learning. It is noteworthy that students did not mention listening skill at all while they mentioned reading and pronunciation as the last two language skills they improved through the use of AI chatbots, similar to their teachers. The improvement in speaking and fluency skills was also found by Duong and Suppasetseree although some drawbacks were also reported by their participants. The reliance on reading and writing more than listening and speaking is once again highlighted here, supporting existing literature.

Participants perceived benefits from AI usage in this study are consistent with their AI utilization strategies. Benefits reported in this study include AI chatbots' time efficiency in material development, simplified feedback delivery, personalized guidance, and the provision of equal opportunities. These results are complimentary to the existing research as they were also elicited by Ozer (2024), Mohamed (2024), and Yuen and Schlote (2024). In addition, Fryer et al. (2019) highlighted chatbots as free and online, which could offer opportunities for language learners globally.

In terms of the challenges with AI chatbot usage in language teaching and learning, our data indicated that an increase in academic dishonesty cases emerged as the primary concern from the teachers' perspectives. Mohamed (2024) similarly identified this as one of the disadvantages of the AI-based chatbot, ChatGPT. Both teachers and students in this study reported deficiencies in language development stemming from an overreliance on AI chatbots. Overreliance on these bots could be a serious issue, especially for students as it would lead them to laziness not only in their language learning but also overall academic studies, a concern echoed previously by Mohamed (2024), Ozer (2024), and Patty (2024). Insufficient student autonomy was one of the concerns reported by teachers in this study, which was also mentioned by Ozer as a lack of awareness about the usage. The teachers emphasized the importance of students being autonomous learners to make the most of AI chatbots and their features in their language learning. Finally, both teachers and students in this study acknowledged their difficulty in crafting appropriate prompts for AI chatbots. This could be partly because of the proficiency levels of the students as mentioned by themselves both in this study and Alrajhi (2024). The students in that study believed a certain level of language proficiency was necessary to interact with the chatbot in the first place. In their systematic review, Jinming and Daniel (2024) similarly reported findings from Chen et al. (2023) that the language proficiency of the learners is likely to influence the topics that they can discuss with chatbots. It is noteworthy to present complementary results from the Turkish context alongside the other distinct contexts in which the aforementioned studies were conducted.

Finally, both teachers and students in this study expressed some expectations for AI to enhance their foreign language teaching and learning. Teacher participants mentioned a need for AI literacy training for both students and teachers while student participants emphasized a need for explicit school wide AI policy. Similar challenges, such as a lack of awareness regarding the usage and a lack of institutional support, were also identified by Ozer (2024). As stated earlier, a significant number of students and teachers worldwide have begun to use AI chatbots in their educational lives without adequate guidance or a solid foundation. Addressing the expectations of foreign language learners and teachers, from their immediate environments to broader contexts, could help mitigate some of these issues.

7. Conclusion

This study has reached several conclusions that future researchers can consider when planning their studies. Among the students and teachers, the most widely used AI chatbot is ChatGPT, with both groups commonly utilizing it to support students' writing skills improvement. While teachers prioritize time efficiency in lesson preparation and material development as the primary advantage of using AI chatbots in their professional work, students view on-the-spot language use suggestions and instant corrections as the most significant benefits. Despite many similarities between the participant groups, there are subtle differences in how they use AI chatbots to enhance student learning. For teachers, the main applications included developing engaging learning activities, facilitating brainstorming for topics, and creating model essays for writing courses. For students, the focus was on creating model essays, gathering and synthesizing information on a topic, and error correction. However, not all feedback from our participants was positive. Teachers reported an increase in academic dishonesty cases as more students employed AI chatbots, and students who became overly dependent on AI chatbots for language production tended to lose confidence in their learning skills and fall behind in language development.

The AI-related challenges identified by students included inaccurate or misleading information, limited language development due to overreliance on AI chatbots, and an inability to craft effective AI prompts. To address issues of overreliance and reduced confidence, educators should scaffold activities that blend AI use with peer and teacher feedback, ensuring that human interaction and reflective thinking remain central to the learning process. Most teachers emphasized the expectation for AI literacy education for both students and teachers. This expectation not only encourages the ethical use of AI-powered tools among students but also provides essential guidance on how to harness AI to enhance their language learning. A clear implication for language educators is the need to integrate AI literacy into language curricula, helping learners understand not just how to use AI tools, but how to do so critically, ethically, and effectively.

This study has three main limitations. First, the participants were selected from a single institution, which may have resulted in a more homogeneous sample, thereby limiting the diversity of perspectives captured in the study.

Second, as the advancements in AI technology have great rapidity, it could be difficult for academic publications to keep up with them. For example, just when we were about to start collecting data for this study, ChatGPT-4o was released on May 13, 2024. Some parts of our data reflect capabilities that were not available in AI chatbots at that time but have since become possible with GPT-4o. Because our participants did not have the opportunity to experience these newer capabilities, their responses were shaped accordingly. Finally, although there is no universally accepted sample size for studies employing qualitative methodologies (Elo et al., 2014), the interpretations of the data in the present chapter may be limited due to the relatively small sample size.

Regarding the implications of this study, we believe that AI chatbots can serve as valuable tools for efficiently searching through vast amounts of online data and this feature will likely remain significant for EFL learners. However, it is crucial for teachers to make students aware that the information retrieved may not always be accurate. Another implication is that ethical academic considerations, especially from educators' perspective, should not be disregarded. Schools should offer both students and teachers training in AI literacy and the responsible use of AI should be promoted for students in particular. Similarly, Baidoo-Anu and Ansah (2023) also refer to an AI-dominated work environment after school arguing that it is important we prepare students by teaching them in the classroom how to use AI tools constructively and safely so that they will be ready for their future professional lives. Furthermore, schools should clearly outline in their AI guidelines which practices are unacceptable, as this clarification could help ensure the ethical use of AI in education. In addition, a balance between technology and human interaction is crucial. While AI chatbots can help students in their language learning process in terms of many aspects, it remains important that these students maintain interpersonal relationships with peers and teachers to meet their human needs without compromising their mental health. Moreover, if teachers leverage AI technologies to reduce their workload in areas such as feedback delivery, test preparation, and administrative tasks, they may create additional time for in-class teaching and foster enhanced, more productive student interactions.

References

Adamopoulou, E., & Moussiades, L. (2020). Chatbots: History, technology, and applications. *Machine Learning with Applications, 2.* https://doi.org/10.1016/j.mlwa.2020.100006

Alrajhi, A. S. (2024). Artificial intelligence pedagogical chatbots as L2 conversational agents. *Cogent Education, 11*(1), 2327789. https://doi.org/10.1080/2331186X.2024.2327789

Baidoo-Anu, D., & Ansah, L. O. (2023). Education in the era of generative artificial intelligence (AI): Understanding the potential benefits of ChatGPT in promoting teaching and learning. *Journal of AI, 7*(1), 52-62. https://doi.org/10.61969/jai.1337500

Barrot, J. S. (2023). Using ChatGPT for second language writing: Pitfalls and potentials. *Assessing Writing, 57,* 1-6. https://doi.org/10.1016/j.asw.2023.100745

Begum, B., Yadav, P. S., Pandey, M., & Reddy, P. R. (2024). Examining the challenges and opportunities of integrating chatbots into educational settings. *Educational Administration: Theory and Practice, 30*(4), 6855-6861. https://doi.org/10.53555/kuey.v30i4.2486

Berry, D. M. (2023). The limits of computation: Joseph Weizenbaum and the ELIZA Chatbot. *Weizenbaum Journal of the Digital Society, 3*(3). https://doi.org/10.34669/WI.WJDS/3.3.2

Bozkurt, A., & Sharma, R. C. (2024). Are we facing an algorithmic renaissance or apocalypse? Generative AI, chatbots, and emerging human-machine interaction in the educational landscape. *Asian Journal of Distance Education, 19*(1), 1-12. https://asianjde.com/ojs/index.php/AsianJDE/article/view/781

Chan, C. K. Y. (2023). A comprehensive AI policy education framework for university teaching and learning. *International Journal of Educational Technology in Higher Education, 20,* 23. https://doi.org/10.1186/s41239-023-00408-3

Chen, H. H. J., Yang, C. T. Y., & Lai, K. K. W. (2023). Investigating college EFL learners' perceptions toward the use of Google Assistant for foreign language learning. *Interactive Learning Environments, 31*(3), 1335-1350. https://doi.org/10.1080/10494820.2020.1833043

Cohen, L., Manion, L., & Morrison, K. (2002). *Research methods in education.* Routledge.

Council of Europe. Council for Cultural Co-operation. Education Committee. Modern Languages Division. (2001). *Common framework of reference for languages: Learning, teaching, assessment.* Cambridge University Press. https://rm.coe.int/common-european-framework-of-reference-for-languages-learning-teaching/16802fc1bf

Dhivvya, J. P., & Karnati, S. B. (2024, March). BuddyBot: AI powered chatbot for enhancing English language learning. In S. Kumar, P. Ranjan, S. Jain, & A. K. Kamal (Eds.), *2024 IEEE International Conference on Interdisciplinary Approaches in Technology and Management for Social Innovation (IATMSI), 2,* 1-6. IEEE. https://doi.org/10.1109/iatmsi60426.2024.10502595

Duong, T., & Suppasetseree, S. (2024). The effects of an artificial intelligence voice chatbot on improving Vietnamese undergraduate students' English-speaking skills. *International Journal of Learning, Teaching and Educational Research, 23*(3), 293-321. https://doi.org/10.26803/ijlter.23.3.15

Elo, S., Kääriäinen, M., Kanste, O., Pölkki, T., Utriainen, K., & Kyngäs, H. (2014). Qualitative content analysis: A focus on trustworthiness. *SAGE Open, 4*(1). https://doi.org/10.1177/2158244014522633

Fitria, T. N. (2021, December). Artificial intelligence (AI) in education: Using AI tools for teaching and learning process. *Proceeding Seminar Nasional & Call for Paper, 4*(1), 134-147.

Fryer, L. K., Ainley, M., Thompson, A., Gibson, A., & Sherlock, Z. (2017). Stimulating and sustaining interest in a language course: An experimental comparison of chatbot and human task partners. *Computers in Human Behavior,* 75, 461-468. https://doi.org/10.1016/j.chb.2017.05.045

Fryer, L. K., Nakao, K., & Thompson, A. (2019). Chatbot learning partners: Connecting learning experiences, interest and competence. *Computers in Human Behavior, 93,* 279-289. https://doi.org/10.1016/j.chb.2018.12.023

Gökçearslan, S., Tosun, C., & Erdemir, Z. G. (2024). Benefits, challenges, and methods of artificial intelligence (AI) chatbots in education: A systematic literature review. *International Journal of Technology in Education, 7*(1), 19-39. https://doi.org/10.46328/ijte.600

Han, J., Yoo, H., Kim, Y., Myung, J., Kim, M., Lim, H., Kim, J., Lee, T. Y., Hong, H., Ahn, S.-Y., & Oh, A. (2023, July). RECIPE: How to integrate ChatGPT into EFL writing education. *Proceedings of the Tenth ACM Conference on Learning @ Scale (L@S '23),* 416-420. https://doi.org/10.1145/3573051.3596200

Haristiani, N. (2019). Artificial intelligence (AI) chatbot as language learning medium: An inquiry. *Journal of Physics: Conference Series,* 1387. http://dx.doi.org/10.1088/1742-6596/1387/1/012020

Jinming, D. U., & Daniel, B. K. (2024). A systematic review of AI-powered chatbots in EFL speaking practice: Transforming language education. *Computers and Education: Artificial Intelligence,* 100239. https://doi.org/10.1016/j.caeai.2024.100230

Kim, N. Y., Cha, Y., & Kim, H. S. (2019). Future English learning: Chatbots and artificial intelligence. *Multimedia-Assisted Language Learning, 22*(3). http://doi.org/10.15702/mall.2019.22.3.32

Korda, D. R., Dapaah, E. O., & Akolgo, E. A. (2024). The role of AI chatbots in education: A review. *Academy Journal of Science and Engineering, 18*(1), 79-87. https://www.ajol.info/index.php/ajse/article/view/291461

Mohamed, A. M. (2024). Exploring the potential of an AI-based chatbot (ChatGPT) in enhancing English as a foreign language (EFL) teaching: Perceptions of EFL faculty members. *Education and Information Technologies, 29*(3), 3195-3217. https://doi.org/10.1007/s10639-023-11917-z

Ozer, O. (2024). AI language models: A breach of academic integrity in online language learning? *Studies in Language Assessment, 13*(1), 237-262. https://doi.org/10.58379/EZYS5680

Patty, J. (2024). The use of AI in language learning: what you need to know. *Jurnal Review Pendidikan dan Pengajaran (JRPP), 7*(1), 642-654. https://doi.org/10.31004/jrpp.v7i1.24609

Punar Özçelik, N., & Yangın Ekşi, G. (2024). Cultivating writing skills: The role of ChatGPT as a learning assistant—a case study. *Smart Learning Environments, 11*(1). https://doi.org/10.1186/s40561-024-00296-8

Shikun, S., Grigoryan, G., Huichun, N., & Harutyunyan, H. (2024). AI chatbots: Developing English language proficiency in EFL classroom. *Arab World English Journal (AWEJ), Special Issue on ChatGPT,* 292-305. https://dx.doi.org/10.24093/awej/ChatGPT.20

Yuen, C. L., & Schlote, N. (2024). Learner experiences of mobile apps and artificial intelligence to support additional language learning in education. *Journal of Educational Technology Systems, 52*(4), 507-525. https://doi.org/10.1177/00472395241238693

Zhang, R., Zou, D., & Cheng, G. (2023). A review of chatbot-assisted learning: pedagogical approaches, implementations, factors leading to effectiveness, theories, and future directions. *Interactive Learning Environments, 32*(8), 4529-4557. https://doi.org/10.1080/10494820.2023.2202704

Zhang, Z., & Huang, X. (2024). The impact of chatbots based on large language models on second language vocabulary acquisition. *Heliyon, 10*(3). https://doi.org/10.1016/j.heliyon.2024.e25370

Chapter 7

Exploring ChatGPT's role in enhancing English speaking proficiency

Xiaoxuan Li
Pennsylvania State University

Abstract

This chapter investigates the potential and limitations of ChatGPT as a tool for enhancing English-speaking proficiency among Chinese learners of English. Drawing on a qualitative narrative inquiry framework, the study centers learners' voices to explore how ChatGPT's text- and voice-based features shape speaking practice, motivation, and confidence. Guided by an affordance theoretical lens, the chapter examines the technological, pedagogical, and social affordances of ChatGPT and how learners perceive its role in their language development. Findings from six participants reveal that while ChatGPT offers accessible, flexible, and low-anxiety speaking opportunities, challenges such as context loss, lack of corrective feedback, and cultural limitations persist. Notably, participants perceived ChatGPT as a responsive and supportive partner, fostering engagement and confidence, particularly through scenario-based conversations. However, overreliance on its interpretive tolerance may reinforce poor grammar habits among lower-proficiency learners. The study concludes by advocating for a blended approach that combines AI-enabled flexibility with structured pedagogical support to maximize the impact of chatbot-assisted speaking practice in English language learning.

Keywords: ChatGPT, English speaking proficiency, Language learning anxiety, AI affordances, Native inquiry

* * *

1. Introduction

Over the past few decades, the advancement of artificial intelligence (AI) technologies has revolutionized traditional teaching methods and profoundly reshaped students' learning experiences (Ou, 2024). Its integration into the ever-changing landscape of education has provided greater accessibility to personalized learning experiences and innovative instructional tools. The convergence of sophisticated algorithms and intelligent machine learning models enables real-time feedback and interactions, fostering a more dynamic and responsive learning environment.

In the realm of second (L2) and foreign language (FL) education, AI technology presents profound potential to go beyond the constraints of traditional classroom settings and address long-standing learner challenges, including insufficient language exposure (Collins & Muñoz, 2016), learner anxiety (El Shazly, 2021), lack of authentic learning contexts (Kim et al., 2021), and limited opportunities for timely dialogic feedback (Huang et al., 2022).

Among AI innovations, ChatGPT, a widely accessible AI chatbot, has emerged as a promising tool for enhancing and transforming language teaching and learning. It has garnered increasing attention from scholars and practitioners for its potential to optimize learning gains in ways that traditional methods may not afford (Fryer et al., 2019). This potential is particularly relevant to English language learning, where speaking proficiency is both highly valued and difficult to achieve (Joo et al., 2019; McKay, 2018). Achieving fluency in speaking is central to effective communication and is often regarded as a key indicator of language competence (Hanzawa & Suzuki, 2023). However, it remains a significant challenge for L2/FL learners due to the limited opportunities they have to practice speaking in authentic language contexts (Khan & Takkac, 2021). ChatGPT, as a dialogue-based system, offers a promising response to these challenges by providing flexible and low-pressure conversational interactions beyond the classroom (Zhou et al., 2023).

However, like other groundbreaking technologies, the incorporation of AI into language educational practices also raises important concerns. Beyond ethical issues and data privacy, researchers have noted several limitations of AI chatbots for language learning, including communication breakdowns (Kim, 2015), limited emotional support and adaptability compared to human interactions (El Shazly, 2021), and restricted responsiveness due to reliance on preset knowledge bases (Hew et al., 2022). Therefore, it is essential to examine both the opportunities and limitations of AI chatbots in English language learning, with particular attention to how ChatGPT supports the development of speaking proficiency.

This chapter aims to explore how Chinese learners of English perceive and experience the use of ChatGPT in developing their speaking proficiency. Drawing on qualitative narrative inquiry with six learners, the chapter investigates both the benefits and constraints of chatbot-assisted speaking practice. By situating learners' voices at the center, it contributes to broader discussions of how AI technology can shape the future of language education while clarifying its current limitations.

2. Literature review

In recent years, the progress and expansion of AI technology have tremendously transformed the educational landscape, with chatbots emerging as one of the most influential innovations. Among them, ChatGPT has emerged as a breakthrough for its advanced natural language processing, gaining rapid popularity for its ability to generate contextually relevant, human-like responses (Ray, 2023). Drawing on extensive text datasets, ChatGPT excels at conversational tasks by generating responses in varied styles and tones while demonstrating strong contextual understanding (Huang et al., 2022; Ray, 2023). These features make ChatGPT a promising tool for English language education, especially in developing speaking proficiency.

English proficiency is widely regarded as essential for global communication and access to educational and professional opportunities (Phillipson, 2013). Researchers often measure the success of language acquisition by an individual's ability to engage in conversations in the target language (Leong & Ahmadi, 2017). However, mastering speaking remains one of the most critical and challenging skills for many language learners. As the ultimate tool for effective communication, speaking involves a meaning-constructing process shaped by the parties involved, their experiences, the context, and the purpose of the interaction. These variables add to the complexity of acquiring speaking skills and often heighten learners' anxiety.

Speaking is consistently identified as a significant source of anxiety in language learning (Kasbi & Shirvan, 2017). Similarly, Horwitz et al. (1986) and Price (1990) found that speaking is more anxiety-provoking than other language skills, an argument reinforced by Bashori et al. (2020), who highlighted that foreign language anxiety predominantly manifests during speaking practices. Several factors contribute to speaking anxiety in language learning, particularly the classroom environment, where learners fear making mistakes or facing negative judgments. Reducing public scrutiny can create a more relaxed context for producing language output—an environment that ChatGPT is well-suited to provide. Studies show that English language learners (ELLs) feel less embarrassed and more motivated to engage in conversations with chatbots (Hadi & Junor, 2022; Tai & Chen, 2022). For instance, Tai and Chen (2022) found that Google Assistant improved communicative confidence and reduced speaking anxiety among ELLs. The human-machine interaction enables learners to express their thoughts more freely, focusing on generating meaningful ideas rather than performing flawless speech. By reducing anxiety, tools like ChatGPT can improve learners' oral communication skills (Ataiefar & Sadighi, 2017).

Research has found a strong link between speaking anxiety and fear of negative evaluation (Tsiplakides & Keramida, 2009). Learners are highly anxious about receiving negative evaluations from their teachers and peers regarding their language errors in speaking. In some cases, direct corrections in speaking activities are perceived as threats to learners' self-esteem rather than as a helpful tool for improvement, reducing their motivation to participate. Demotivation is particularly evident in formal language learning settings, where the compulsory status of language learning and high-stakes examinations can further discourage engagement (Graddol, 2006). Ross and Rivers (2018) similarly observed that college students experienced more intense emotions in classrooms than in informal contexts, and that enjoyment increased when language learning could be applied in real-world situations. Interacting with ChatGPT allows learners to practice speaking without feeling judged negatively while still receiving instant feedback on grammar and vocabulary (Karataş et al., 2024). Language learners often find interactions mediated by chatbots to be enjoyable, motivating, and engaging (Xiao et al., 2023).

In addition, students often lack sufficient opportunities to practice speaking in L2/FL classrooms, which are widely recognized as constrained by time limitations (Collins & Muñoz, 2016). As a productive skill, speaking requires two key components: accuracy, the correct use of vocabulary, grammar, and pronunciation; and fluency, the ability to sustain smooth and spontaneous speech (Derakhshan et al., 2016). Because formal learning classrooms often fall short in providing extended speaking practice and interaction, growing attention has turned toward alternative environments that facilitate language speaking.

Technological advancements have extended language learning opportunities beyond the classroom. The mobility, flexibility, and accessibility of AI chatbots provide an advantage in overcoming time and location constraints when practicing speaking skills. The extensive datasets on which AI chatbots rely provide unlimited access to diverse conversational scenarios, facilitating varied conversation practices (Bansal et al., 2024). However, their effectiveness depends on learners' self-directed engagement, as autonomy is essential for deep and meaningful learning. ChatGPT, in particular, has demonstrated its ability to enhance self-directed language learning by generating contextually relevant responses and fostering meaningful conversations (Li et al., 2024).

While ChatGPT offers vast potential for language learning, it is also critical to recognize its limitations in supporting speaking proficiency. One primary concern is the potential for cultural biases and stereotypes, as ChatGPT was trained on large datasets that often privilege Western perspectives (Grassini, 2023). Its output may lack diverse representations of gender, race, ethnicity, and cultural backgrounds (Kartal, 2023). For example, Zheng and Stewart (2024) found that WEIRD (Western, educated, industrial, rich, and democratic) cultural values dominated ChatGPT-delivered moral dilemmatic narratives, which could perpetuate the longstanding influence of Western intellectual and cultural production in English language education (Phillipson, 2013). For ELLs unfamiliar with these cultural nuances, critically evaluating ChatGPT-generated materials can be difficult. Since speaking proficiency involves not only grammar and pronunciation but also pragmatic and intercultural competence (Usó-Juan & Martínez-Flor, 2008), ChatGPT's limited cultural diversity may prevent it from fully meeting learners' communicative needs.

Despite substantial progress in generating coherent responses, there is still room for improvement in maintaining context during extended, multi-turn conversations (Ray, 2023). In lengthy dialogues, ChatGPT may lose track of complex contexts, resulting in responses that are irrelevant or unpredictable and disrupt the flow of conversation. Such issues can be particularly challenging for lower-proficiency learners, who may struggle to follow the conversation and generate meaningful responses. Communication breakdown may also arise from learners' mispronunciations and sudden topic shifts (Kim et al., 2021). Another limitation is ChatGPT's inability to interpret nonverbal cues such as body language, gestures, and facial expressions. It also fails to grasp nuanced human speech elements such as "humor, pun, irony, and sarcasm" (Mukherjee, 2020, p. 91). These gaps can leave learners feeling frustrated or embarrassed, which may discourage them from continuing to practice.

While existing research highlights various advantages and limitations of AI chatbots in L2/FL language education, there are notable gaps that this study intends to address. Despite the substantial body of studies examining the effectiveness of AI chatbots in language learning, there is a scarcity of qualitative studies that delve into language learners' personal experiences, perceptions, and attitudes. In a systematic review of 24 studies on chatbots and English oral skills, only two used qualitative methods, reflecting a predominant reliance on mixed or quantitative approaches (Du & Daniel, 2024). Understanding learners' first-hand perspectives is essential, as it reveals how ChatGPT's interactive features align with their needs, preferences, and challenges (Jeon & Lee, 2023; Ulla et al., 2023; Xiao & Zhi, 2023).

A second gap concerns the under-exploration of ChatGPT's effectiveness in improving speaking proficiency specifically (Li et al., 2024; Muniandy & Selvanathan, 2024). ChatGPT's text-centric interface lacks speech affordances for language learners (Kostka & Toncelli, 2023). With the introduction of speech capabilities in September 2023, new opportunities for oral interaction have emerged, yet these remain largely unexplored in the literature. Because speaking is a multifaceted skill requiring not only accurate sounds but also grammar, cultural awareness, and conversational strategies (Derakhshan et al., 2016), it is essential to investigate how learners engage with ChatGPT's text- and voice-based interactions.

This current study addresses these gaps by collecting rich and in-depth qualitative narratives that capture the nuances behind language learning. With recent advancements enabling ChatGPT to process speech and auditory input, this study also investigates how language learners interact with both the text and voice capabilities. The learners' narratives offer a comprehensive and critical understanding of the potential of ChatGPT to enhance oral language skills, particularly in terms of fluency, accuracy, and conversational competence.

3. The study

This qualitative case study examines the effectiveness of ChatGPT in enhancing English language speaking proficiency among Chinese learners. With the recent introduction of voice capabilities in ChatGPT, it is essential to explore how both its text-based and voice-based features influence Chinese learners' English-speaking skills. The primary objective of this study is to examine the perceived benefits and limitations of using ChatGPT as a novel tool for enhancing L2 speaking proficiency, with a focus on learners' own perspectives. By collecting individual narratives, this study aims to explore how ChatGPT's interactive capabilities impact learners' confidence, motivation, and proficiency in speaking English. The findings also provide insights into effective and practical strategies for integrating ChatGPT into L2/FL education while highlighting potential challenges that may arise in practice. These insights aim to guide learners, educators, and researchers in leveraging AI tools more effectively for language teaching and learning. Two research questions guided this study: (1) How do Chinese ELLs at different proficiency levels experience and perceive the effectiveness of ChatGPT in developing their speaking skills and (2) What challenges do Chinese ELLs face when using ChatGPT as a tool to practice English speaking?

3.1 Theoretical framework

Grounded in Kirschner et al.'s (2004) affordance theoretical framework, this study examines the technological, pedagogical, and social affordances associated with ChatGPT in language learning. The concept of affordance, first introduced by Gibson (1986) in ecological psychology, refers to the relationship between environmental properties and an organism's capacity for action.

Building on this foundation, scholars across disciplines have adapted the term to describe context-specific opportunities for interaction.

In the field of language learning, affordances represent the possibilities for language acquisition that are co-constructed by the properties of the environment and the learner's perceptions, motivations, and willingness to engage. This perspective emphasizes the integration of learners' prior learning experiences, future expectations, and subjective evaluations in shaping language learning outcomes.

Within this framework, technological affordances concern the usability and utility of ChatGPT as a learning tool. In this study, these affordances refer to the design features of the platform that facilitate how easily and effectively learners can utilize ChatGPT for speaking practice. These design features include the user-friendliness of the platform (usability) and its effectiveness in meeting learners' needs for practicing and enhancing their speaking skills (utility).

Pedagogical affordances pertain to how a learning tool shapes whether and how specific learning behaviors can occur within a particular setting. Understanding how the features of a learning tool promote desired outcomes while potentially imposing constraints is essential. For example, ChatGPT's capability to generate conversational prompts and provide relevant feedback may facilitate speaking practice in diverse scenarios. At the same time, it may limit opportunities for meaningful learning if feedback lacks depth or becomes repetitive.

Lastly, social affordances refer to how a tool facilitates social interactions among participants, to strengthen group cohesion and a sense of belonging (Kirschner et al., 2004). In this study, the researchers determined social affordances primarily by examining ChatGPT's capacity to support social presence among learners. Garrison (2011) defined social presence as learners' ability to identify with a group, engage meaningfully in a supportive environment, and gradually build personal and affective relationships. ChatGPT's ability to adapt to varied communication styles and provide empathetic, personalized feedback helps cultivate a sense of connectedness and belonging. Social affordances emerge when learners perceive ChatGPT as more than a technological tool, but as a responsive conversational partner capable of engaging in meaningful interactions.

3.2 Research methodology

This study employs qualitative narrative inquiry to delve into the experiences and perceptions of Chinese ELLs who use ChatGPT to enhance their English-speaking skills. Narrative inquiry, often described as the study of experience as story (Connelly & Clandinin, 2006), enables researchers to explore and document individuals' beliefs and experiences in rich detail (Bell, 2002). Rooted in anthropology, sociology, and psychology, narrative inquiry extends beyond merely telling and collecting stories; it also involves analyzing and interpreting them. It entails a comprehensive analysis of narrative data through multiple lenses, incorporating the perspectives of research participants, the researcher, and the broader sociocultural and institutional contexts that inform their experiences.

Researchers have extensively used narrative inquiry as a research methodology in the field of language teaching and learning. By capturing the emotional, cognitive, and social dimensions of language acquisition, rich personal accounts provide insight into how learners navigate the dynamics of their language learning journeys.

The focus of this study was to understand how Chinese ELLs perceive their interactions with ChatGPT, including both the challenges they face and the successes they achieve. These learners, who have traveled to the United States to learn English as a second language, regularly engage with ChatGPT outside formal classroom settings as part of their daily learning routines. Because learners closely connect second and foreign language learning with their broader life experiences, and because they typically engage in ChatGPT-based speaking practice outside formal classrooms, researchers consider narrative inquiry an ideal approach for capturing these experiences.

Although the starting point of narrative inquiry is individuals' stories, it also delves into the social, cultural, and institutional narratives within which individuals' experiences are embedded and shaped (Clandinin & Rosiek, 2006). Participants' experiences with ChatGPT are shaped not only by their efforts but also by interactions with the broader community, reflecting underlying interpersonal and social dynamics. This holistic approach enables a comprehensive understanding of how ChatGPT is utilized and perceived by integrating personal narratives within broader sociocultural frameworks.

3.3. Data collection and data analysis

Data for this study was collected through semi-structured, in-depth interviews that explored participants' experiences and perceptions of using ChatGPT to improve their English-speaking proficiency. Six participants, including four females and two males, were recruited through purposive sampling from an Intensive English Program (IEP) affiliated with a university in the Southwestern United States. IEPs provide an ideal context for this study, as they are designed to support international students' transition into U.S. higher education by offering intensive training in the four core language skills (listening, speaking, reading, and writing) across different proficiency levels.

All participants met the following criteria: (a) Chinese nationality; (b) at least two years of studying English as a second language in the U.S.; (c) purposeful use of ChatGPT to practice English speaking for at least one year, through text-based features, voice-based features, or both.

The researcher categorized participants' English proficiency levels according to the program's achievement scale: Jessica and John were at the elementary level, Dylan and Chloe at the intermediate level, and Ella and Lucy at the advanced level.

During recruitment, potential participants completed a questionnaire that captured detailed information about their use of ChatGPT for English speaking practice, including practice frequency, average session length, and total weekly hours. All selected participants reported practicing with ChatGPT at least three times per week, averaging 45 minutes per session and approximately 3–4 hours per week overall. This frequency ensured that participants had regular and sufficient interaction with ChatGPT, allowing them to observe a meaningful impact over time.

Chinese students constitute the largest group of international students in the U.S. (Su et al., 2021). They face a pressing need to improve English-speaking skills as English shifts from an FL in China to an L2 in the U.S. context. In China, schools and teachers primarily focus English learning on passing exams and meeting academic requirements, emphasizing reading and writing over speaking and listening (Pan, 2015). This so-called "dumb and deaf" approach often leaves students with the least confidence in their speaking abilities (Liu et al., 2016). Given these realities, ChatGPT presents a potentially valuable tool by offering additional interactions, real-time feedback, and a flexible and supportive environment for conversational practice.

The researcher analyzed the data concurrently with its collection, consistent with the interactive and iterative nature of qualitative research (Miles et al., 2020). The process involved transcribing interviews, organizing field notes, and coding using NVivo software to identify key themes and patterns. The researchers developed key themes (e.g., technological limitations, increased confidence, and psychological frustrations) through an iterative process of reviewing and refining the data. The findings were then synthesized into a narrative, supported by representative participant quotes, offering nuanced insights into both the benefits and challenges that Chinese ELLs experienced when using ChatGPT to practice English speaking.

4. Results

The participants' narratives reveal nuanced and critical insights into how Chinese ELLs at different English proficiency levels engage with ChatGPT to develop speaking skills. All participants demonstrated a less frequent use of the voice-based feature compared to the text-based feature of ChatGPT, a tendency that was especially pronounced among elementary-level learners. For these participants, ongoing pronunciation struggles frequently led to voice recognition errors, resulting in frustration and heightened self-consciousness about their limited fluency. By contrast, text-based interactions were perceived as more accommodating and less stressful, allowing for additional processing time to ensure their responses were well-articulated. This extra time was particularly valuable in supporting clearer and more confident speaking.

In contrast, learners at the intermediate and advanced levels favored text-based features for different reasons. They preferred using voice chat without others present, which was not always feasible in public spaces like libraries or coffee shops. The need for a private setting highlights the importance of a controlled environment for effective speaking practice. Text-based interactions provide a solution for learners to engage in practice asynchronously or in environments where voice communication may be too stressful. As Chloe explained, "Speaking out loud, even to a machine, just makes me panic, especially when there are others around." Such anxiety is common, as speaking in front of others often heightens self-consciousness (Chen, 2022). Written communication helps mitigate feelings of self-consciousness and anxiety, a finding also supported by Namaziandost et al. (2021), who found that Iranian ELLs experienced lower anxiety levels in synchronous computer-mediated text-chat groups compared to voice chat groups.

Participants reported overall positive experiences of using ChatGPT to improve their speaking skills. ChatGPT's accessibility and flexibility were frequently noted in the participants' narratives. They valued its immediate availability, real-time assistance, and capacity to adapt to individual learning needs. The participants viewed ChatGPT's capacity to stimulate diverse conversational topics and scenarios as a significant advantage over their previous classroom experiences, which often relied on repetitive exercises. Accessibility was essential for these students, who faced challenges securing consistent language support while abroad. Ella, who was highly committed to improving her speaking, noted, "I wanted to practice English with my friends or tutor back in China, but they are usually asleep when I need to practice because of the time difference, you know." The social isolation that international students experience while studying abroad can be alleviated through the availability of technological tools.

Participants also emphasized the social affordances of ChatGPT, describing it as a language tutor, conversational partner, and learning assistant. Interestingly, none explicitly identified ChatGPT as an intelligent machine. Instead, they attributed human-like qualities to it and saw it as a supportive presence in their learning journey. As Jessica shared, "ChatGPT feels like a loyal and reliable friend who is always there whenever I need to practice or have any questions; it never gets tired of my endless questions, so I can bother it anytime." The participants' tendency to humanize ChatGPT underscores its emotional support capabilities, contributing to a more enjoyable and engaging practice experience. Moreover, participants particularly valued its empathetic and varied responses, which created a more authentic conversational expertise. These perceptions align with Stöckli et al.'s (2024) finding that ChatGPT-4's ability to display personality traits enhances user satisfaction.

In terms of pedagogical affordances, learners valued ChatGPT more for its capability to support the mastery of language functions and sociocultural norms than for its role in enhancing the mechanics of English speaking, such as grammar, vocabulary, and sentence structures. Despite relying primarily on text chats, four participants noted ChatGPT's effectiveness in helping them navigate extended dialogues with native speakers. ChatGPT's ability to generate scenario-based conversations, such as mailing items, making doctor appointments, or ordering meals, helped them practice negotiating arguments, offering advice, and expressing opinions. For example, Dylan explained that he often asked ChatGPT to create detailed conversation scripts for different scenarios before real-life interactions. By rehearsing these scripts, he became familiar with typical language patterns, common phrases, and response strategies expected in those situations. Lucy similarly mentioned, "When I speak with native speakers, like in restaurants, I know I cannot speak perfect English, but I feel more confident to engage because I know what to expect and how to respond." Moreover, three participants emphasized that ChatGPT expanded their understanding of cultural aspects often overlooked in classroom instruction. They enjoyed exploring cultural topics, such as holidays and social customs, in a safe space, free from the fear of making culturally insensitive mistakes.

The most frequently reported challenge was ChatGPT's tendency to generate out-of-context responses, leading to communication breakdowns. This issue often occurred when conversations shifted topics or when prompts contained multiple topics simultaneously. The frequent occurrence of irrelevant responses caused frustration among participants, as they had to modify and clarify their prompts continually. As John explained, "Sometimes I keep getting responses that don't fit what I am talking about, so I end up spending more time editing the prompts than actually practicing English." Learners perceived this inefficiency as a barrier to meaningful speaking practice, as it disrupted conversational flow and distracted their attention from the intended topic. Managing and correcting miscommunications required sustained effort, increasing learners' cognitive load. Instead of focusing solely on language practice, learners had to troubleshoot ChatGPT's responses and modify their prompts. Interactions with ChatGPT required additional mental effort, as learners had to balance both the conversation content and technical issues. For learners seeking smooth and fluid conversation practice, this compromised ChatGPT's efficiency and authenticity as a conversational partner.

Another issue reported by participants at the elementary level was that ChatGPT often responded to fragmented input or minimal grammar without providing corrective feedback, due to its high level of intelligence. The tool was still able to generate relevant responses even when learners' input was incomplete or grammatically inaccurate. Over time, learners began to rely on producing incomplete sentences and neglecting grammatical precision in terms of verb agreement, word order, and overall cohesion. While such tolerance for errors may be beneficial for lowering anxiety and boosting engagement, it also carries unintended consequences, such as reinforcing inaccurate language use. As John pointed out, "It's making me lazy with my grammar. I mean, it can understand me anyway." While the lack of clarity and grammatical accuracy can be acceptable in human-AI interactions, it poses challenges for language learners when transitioning to more formal or nuanced communication situations. Relying on AI's interpretive capabilities to process imprecise input may unintentionally reinforce habits that hinder effective language use. This risk is particularly acute for entry-level learners who heavily rely on explicit feedback to build foundational skills.

This heightened risk suggests the need for structured pedagogical interventions, such as pairing ChatGPT-assisted practice with teacher-led feedback sessions in which learners critically review their outputs and identify recurring grammatical errors. This hybrid approach ensures that gains in confidence and engagement do not come at the expense of long-term linguistic accuracy.

5. Researcher reflexivity

The researcher's identity, professional background, and academic journey inevitably shape the research process and its results. In qualitative inquiry, researchers should purposefully acknowledge their subjectivity, as both their identities and those of the participants influence the co-construction of knowledge (Bourke, 2014).

My extensive experience working with Chinese ELLs affords an insider's understanding of the complexities involved in mastering English-speaking proficiency, particularly the cultural, linguistic, and psychological challenges these learners face. Moreover, sharing a cultural background and lived experience as an ELL enabled me to establish rapport and foster meaningful discussions with participants, particularly around sensitive issues such as self-confidence and language anxiety.

At the same time, my prior involvement with high-proficiency learners and my generally positive experiences with ChatGPT may predispose me to emphasize its benefits and overlook its limitations for learners at other proficiency levels. To address this, I engaged in ongoing self-reflexivity throughout the study, critically examining my own assumptions about AI's role in language learning and questioning how they might influence interpretation. By recognizing and addressing both supportive and critical perspectives, I aimed to present a balanced view of the benefits and limitations of ChatGPT in enhancing English-speaking proficiency.

6. Conclusion

This chapter has examined the effectiveness of ChatGPT in supporting English-speaking skills through a qualitative narrative inquiry involving six Chinese learners. The findings shed light on how ELLs at different proficiency levels engaged with ChatGPT, the benefits they perceived, and the challenges they encountered.

Learners across different proficiency levels predominantly preferred text-based interactions over voice chat when practicing speaking with ChatGPT. For elementary-level learners, the text-based features provided sufficient time to articulate their responses more clearly, while intermediate and advanced learners appreciated the privacy and reduced anxiety these features offered. Overall, participants expressed favorable perceptions of ChatGPT, emphasizing its role as a supportive resource in their language learning. Its technological, pedagogical, and social affordances offered unique opportunities to practice speaking in ways that felt authentic and emotionally engaging, bridging gaps between classroom instruction and real-life interactions.

With its capacity to simulate diverse conversational scenarios and respond empathetically, learners tended to perceive ChatGPT as a reliable conversational partner rather than a mechanical tool. This personalized approach was particularly valuable for learners in isolated environments, such as those studying abroad, by offering a sense of connection and engagement. In addition, participants highlighted ChatGPT's usefulness in familiarizing them with the functional and sociocultural dimensions of language use, rather than focusing solely on mechanical aspects, such as vocabulary and grammar. This finding highlights the persistent disconnection between formal classroom instruction and real-world communication, underscoring the need to integrate context-rich and emotionally engaging experiences into language learning.

However, several challenges arose during participants' use of ChatGPT. Misaligned or irrelevant responses were a common issue, especially when conversations shifted topics or prompts contained multiple ideas. Irrelevant responses caused frustration and required learners to invest additional effort in managing out-of-context responses and refining prompts for better clarity. Another key challenge for elementary-level learners was ChatGPT's tendency to respond to incomplete or inaccurate language without correction. Overreliance on ChatGPT's interpretive capabilities led to the development of harmful language habits, such as producing ungrammatical or incomplete sentences, which could potentially limit the effectiveness of learners' language output.

Taken together, the findings suggest that AI-assisted practice should be balanced with traditional learning methods that reinforce grammatical accuracy and structured speaking drills. A blended model that merges the flexibility of AI tools with the rigor of human instruction offers more substantial potential for sustained language development.

For language teacher educators, learners' narratives demonstrate that ChatGPT can be scaled as a supplementary tool for supporting language speaking, particularly in contexts where learners face heightened speaking anxiety. Notably, ChatGPT's value extends beyond technical skill development, providing social and emotional support that can enhance the effectiveness and sustainability of language learning.

Teachers should prepare learners to engage critically with ChatGPT, not only guiding students in how to use the tool effectively but also helping them develop strategies to avoid overreliance on its interpretive flexibility or tolerance for errors. In practice, this might involve pairing chatbot interactions with targeted feedback sessions, reflection journals, and peer discussions that encourage learners to monitor their language production and critically analyze their progress.

The need to prepare learners to use ChatGPT effectively and critically suggests that teacher education programs should integrate training in human-AI co-design, equipping teachers with the necessary skills to maximize ChatGPT's affordances while remaining attentive to its limitations. Instead of positioning AI as a standalone solution, educators should intentionally integrate its applications into comprehensive pedagogical frameworks that accommodate learners of diverse backgrounds and proficiency levels.

References

Ataiefar, F., & Sadighi, F. (2017). Lowering foreign language anxiety through technology: A case of Iranian EFL sophomore students. *English Literature and*

Language Review, 3(4), 23–34. http://www.arpgweb.com/pdf-files/ellr3(4)23-34.pdf

Bansal, G., Chamola, V., Hussain, A., Guizani, M., & Niyato, D. (2024). Transforming conversations with AI—A comprehensive study of ChatGPT. *Cognitive Computation, 16*(5), 1-24. https://doi.org/10.1007/s12559-023-10236-2

Bashori, M., Van Hout, R., Strik, H., & Cucchiarini, C. (2020). Web-based language learning and speaking anxiety. *Computer Assisted Language Learning, 35*(5–6), 1058–1089. https://doi.org/10.1080/09588221.2020.1770293

Bell, J. S. (2002). Narrative inquiry: More than just telling stories. *TESOL Quarterly, 36*(2), 207. https://doi.org/10.2307/3588331

Bourke, B. (2014). Positionality: Reflecting on the research process. *The Qualitative Report, 19*(33), 1-9. https://doi.org/10.46743/2160-3715/2014.1026

Chen, Y. (2022). Effects of technology-enhanced language learning on reducing EFL learners' public speaking anxiety. *Computer Assisted Language Learning, 37*(4), 789-813. https://doi.org/10.1080/09588221.2022.2055083

Clandinin, D. J., & Rosiek, J. (2006). Mapping a landscape of narrative inquiry: Borderland spaces and tensions. In D. J. Clandinin (Ed.), *Handbook of narrative inquiry: Mapping a methodology* (pp. 35–75). Sage. https://doi.org/10.4135/9781452226552.n2

Collins, L., & Muñoz, C. (2016). The foreign language classroom: Current perspectives and future considerations. *Modern Language Journal, 100,* 133–147. https://doi.org/10.1111/modl.12305

Connelly, F. M., & Clandinin, D. J. (2006). Narrative inquiry. In J. L. Green, G. Camilli, & P. B. Elmore (Eds.), *Handbook of complementary methods in education research* (pp. 477–487). Lawrence Erlbaum Associates.

Derakhshan, A., Khalili, A. N., & Beheshti, F. (2016). Developing EFL learner's speaking ability, accuracy and fluency. *English Language and Literature Studies, 6*(2), 177. https://doi.org/10.5539/ells.v6n2p177

Du, J., & Daniel, B. K. (2024). Transforming language education: A systematic review of AI-powered chatbots for English as a foreign language speaking practice. *Computers and Education: Artificial Intelligence, 6.* https://doi.org/10.1016/j.caeai.2024.100230

El Shazly, R. (2021). Effects of artificial intelligence on English speaking anxiety and speaking performance: A case study. *Expert Systems, 38*(3). https://doi.org/10.1111/exsy.12667

Fryer, L. K., Nakao, K., & Thompson, A. (2019). Chatbot learning partners: Connecting learning experiences, interest and competence. *Computers in Human Behavior, 93,* 279–289. https://doi.org/10.1016/j.chb.2018.12.023

Garrison, D. R. (2011). *E-learning in the 21st century: A framework for research and practice* (2nd ed.). Routledge. https://dl.acm.org/citation.cfm?id=1996369

Gibson, J. J. (1986). *The ecological approach to visual perception.* Lawrence Earlbaum. https://doi.org/10.4324/9781315740218

Graddol, D. (2006). *English next: Why global English may mean the end of 'English as a foreign language.'* The British Council.

Grassini, S. (2023). Shaping the future of education: Exploring the potential and consequences of AI and ChatGPT in educational settings. *Education Sciences, 13*(7), 692. https://doi.org/10.3390/educsci13070692

Hadi, M. S., & Junor, R. S. (2022). Speaking to devices: Can we use Google Assistant to foster students' speaking skills? *Journal of Languages and Language Teaching, 10*(4), 570. https://doi.org/10.33394/jollt.v10i4.5808

Hanzawa, K., & Suzuki, Y. (2023). Does automaticity in lexical and grammatical processing predict utterance fluency development? A six-month longitudinal study in Japanese EFL context. *Journal of Second Language Studies, 6*(2), 290–318. https://doi.org/10.1075/jsls.22007.han

Hew, K. F., Huang, W., Du, J., & Jia, C. (2022). Using chatbots to support student goal setting and social presence in fully online activities: Learner engagement and perceptions. *Journal of Computing in Higher Education, 35*(1), 40–68. https://doi.org/10.1007/s12528-022-09338-x

Horwitz, E. K., Horwitz, M. B., & Cope, J. (1986). Foreign language classroom anxiety. *Modern Language Journal, 70*(2), 125–132. https://doi.org/10.1111/j.1540-4781.1986.tb05256.x

Huang, W., Hew, K. F., & Fryer, L. K. (2022). Chatbots for language learning—Are they really useful? A systematic review of chatbot-supported language learning. *Journal of Computer Assisted Learning, 38*(1), 237–257. https://doi.org/10.1111/jcal.12610

Jeon, J., & Lee, S. (2023). Large language models in education: A focus on the complementary relationship between human teachers and ChatGPT. *Education and Information Technologies, 28*(12), 15873-15892. https://doi.org/10.1007/s10639-023-11834-1

Joo, S. J., Chik, A., & Djonov, E. (2019). The construction of English as a global language in Korean EFL textbooks for primary school children. *Asian Englishes, 22*(1), 68–84. https://doi.org/10.1080/13488678.2019.1627636

Karataş, F., Abedi, F. Y., Gunyel, F. O., Karadeniz, D., & Kuzgun, Y. (2024). Incorporating AI in foreign language education: An investigation into ChatGPT's effect on foreign language learners. *Education and Information Technologies, 29*, 19343-19366. https://doi.org/10.1007/s10639-024-12574-6

Kartal, G. (2023). Contemporary language teaching and learning with ChatGPT. *Contemporary Research in Language and Linguistics, 1*(1), 59–70. https://doi.org/10.62601/crll.v1i1.10

Kasbi, S., & Shirvan, M. E. (2017). Ecological understanding of foreign language speaking anxiety: Emerging patterns and dynamic systems. *Asian-Pacific Journal of Second and Foreign Language Education, 2*(2). https://doi.org/10.1186/s40862-017-0026-y

Khan, S. S., & Takkac, M. (2021). Motivational factors for learning English as a second language acquisition in Canada. *Higher Education Studies, 11*(1), 160. https://doi.org/10.5539/hes.v11n1p160

Kim, B. (2015). The English fever in South Korea: Focusing on the problem of early English education. *Journal of Education & Social Policy, 2*(2), 117–124. http://jespnet.com/journals/Vol_2_No_2_June_2015/15.pdf

Kim, H. S., Cha, Y., & Kim, N. Y. (2021). Effects of AI chatbots on EFL students' communication skills. *Korea Journal of English Language and Linguistics, 21*, 712–734. https://doi.org/10.15738/kjell.21.202108.712

Kirschner, P., Strijbos, J., Kreijns, K., & Beers, P. J. (2004). Designing electronic collaborative learning environments. *Educational Technology Research and Development, 52*(3), 47–66. https://doi.org/10.1007/bf02504675

Kostka, I., & Toncelli, R. (2023). Exploring applications of ChatGPT to English language teaching: Opportunities, challenges, and recommendations. *TESL-EJ, 27*(3). https://doi.org/10.55593/ej.27107int

Leong, L., & Ahmadi, S. M. (2017). An analysis of factors influencing learners' English-speaking skill. *International Journal of Research in English Education, 2*(1), 34–41. https://doi.org/10.18869/acadpub.ijree.2.1.34

Li, B., Wang, C., Bonk, C. J., & Kou, X. (2024). Exploring inventions in self-directed language learning with Generative AI: Implementations and perspectives of YouTube content creators. *TechTrends, 68*(4), 803–819. https://doi.org/10.1007/s11528-024-00960-3

Liu, N., Lin, C., & Wiley, T. G. (2016). Learner views on English and English language teaching in China. *International Multilingual Research Journal, 10*(2), 137–157. https://doi.org/10.1080/19313152.2016.1147308

McKay, S. L. (2018). English as an international language: What it is and what it means for pedagogy. *RELC Journal, 49*(1), 9–23. https://doi.org/10.1177/0033688217738817

Miles, M. B., Huberman, M. A., & Saldana, J. (2020). *Qualitative data analysis: A methods sourcebook (4th ed.).* Sage.

Mukherjee, R. (2020). The introduction of artificial intelligence in ESL teaching: Possibilities and limitations. In *Changing trends in human thoughts and perspectives: Science, humanities and culture* (Part I, pp. 89–93). Jogamaya Devi College.

Muniandy, J., & Selvanathan, M. (2024). ChatGPT, a partnering tool to improve ESL learners' speaking skills: Case study in a public university, Malaysia. *Teaching Public Administration, 43*(1), 4-20. https://doi.org/10.1177/01447394241230152

Namaziandost, E., Razmi, M. H., Hernández, R. M., Ocaña-Fernández, Y., & Khabir, M. (2021). Synchronous CMC text chat versus synchronous CMC voice chat: Impacts on EFL learners' oral proficiency and anxiety. *Journal of Research on Technology in Education, 54*(4), 599–616. https://doi.org/10.1080/15391523.2021.1906362

Ou, S. (2024). Transforming education: The evolving role of artificial intelligence in the students academic performance. *International Journal of Education and Humanities, 13*(2), 163– 173. https://doi.org/10.54097/cc1x7r95

Pan, L. (2015). *English as a global language in China: Deconstructing the ideological discourses of English in language education.* Springer.

Phillipson, R. (2013). *Linguistic imperialism continued.* Taylor & Francis. https://doi.org/10.4324/9780203857175

Price, M. L. (1990). The Subjective experience of foreign language anxiety: Interviews with highly anxious students. In E. K. Horwitz & D. J. Young (Eds.), *Language anxiety* (pp. 101–108). Prentice Hall.

Ray, P. P. (2023). ChatGPT: A comprehensive review on background, applications, key challenges, bias, ethics, limitations and future scope.

Internet of Things and Cyber Physical Systems, 3, 121–154. https://doi.org/10.1016/j.iotcps.2023.04.003

Ross, A. S., & Rivers, D. J. (2018). Emotional experiences beyond the classroom: Interactions with the social world. *Studies in Second Language Learning and Teaching, 8*(1), 103-126. https://doi.org/10.14746/ssllt.2018.8.1.5

Stöckli, L., Joho, L., Lehner, F., & Hanne, T. (2024). The personification of ChatGPT (GPT-4): Understanding its personality and adaptability. *Information, 15*(6), 300. https://doi.org/10.3390/info15060300

Su, Z., McDonnell, D., Shi, F., Liang, B., Li, X., Wen, J., Cai, Y., Xiang, Y., & Yang, L. (2021). Chinese international students in the United States: The interplay of students' acculturative stress, academic standing, and quality of life. *Frontiers in Psychology, 12.* https://doi.org/10.3389/fpsyg.2021.625863

Tai, T., & Chen, H. H. (2022). The impact of intelligent personal assistants on adolescent EFL learners' listening comprehension. *Computer Assisted Language Learning, 37*(3), 1–28. https://doi.org/10.1080/09588221.2022.2040536

Tsiplakides, I., & Keramida, A. (2009). Helping students overcome foreign language speaking anxiety in the English classroom: Theoretical issues and practical recommendations. *International Education Studies, 2*(4). https://doi.org/10.5539/ies.v2n4p39

Ulla, M. B., Perales, W. F., & Busbus, S. O. (2023). 'To generate or stop generating response': Exploring EFL teachers' perspectives on ChatGPT in English language teaching in Thailand. *Learning Research and Practice, 9*(2), 168–182. https://doi.org/10.1080/23735082.2023.2257252

Usó-Juan, E., & Martínez-Flor, A. (2008). Teaching intercultural communicative competence through the four skills. *Revista Alicantina De Estudios Ingleses, 21,* 157-170. https://doi.org/10.14198/raei.2008.21.09

Xiao, F., Zhao, P., Sha, H., Yang, D., & Warschauer, M. (2023). Conversational agents in language learning. *Journal of China Computer-Assisted Language Learning, 4*(2), 300-325. https://doi.org/10.1515/jccall-2022-0032

Xiao, Y., & Zhi, Y. (2023). An exploratory study of EFL learners' use of ChatGPT for language learning tasks: Experience and perceptions. *Languages, 8*(3), 212. https://doi.org/10.3390/languages8030212

Zheng, Y., & Stewart, N. (2024). Improving EFL students' cultural awareness: Reframing moral dilemmatic stories with ChatGPT. *Computers and Education Artificial Intelligence, 6.* https://doi.org/10.1016/j.caeai.2024.100223

Zhou, J., Ke, P., Qiu, X., Huang, M., & Zhang, J. (2023). ChatGPT: potential, prospects, and limitations. *Frontiers of Information Technology & Electronic Engineering, 25*(1), 6–11. https://doi.org/10.1631/fitee.2300089

Chapter 8

Faculty perspectives on adaptive learning in foreign language courses

Shelly Wyatt
University of Central Florida

Baiyun Chen
University of Central Florida

Jessica Tojo-Raible
University of Central Florida

Abstract

This chapter explores the experiences of faculty in the Modern Languages and Literatures department at the University of Central Florida as they integrated personalized adaptive learning (PAL) into their courses. This chapter also provides insights into best practices for implementing adaptive learning technologies in higher education settings, highlighting PAL's capacity to tailor instruction to individual student needs through data-driven analytics, real-time assessment, and machine learning. Faculty participants reported that incorporating PAL into their language courses enhanced student engagement, motivation, and retention by offering customized learning pathways. However, challenges such as time-intensive curriculum development, platform complexity, and cost concerns were also noted. The effectiveness of PAL depends on faculty training, institutional support, and accessible technology solutions. This chapter offers practical insights for educators and administrators seeking to implement adaptive learning, emphasizing faculty collaboration, open educational resources integration, and sustained institutional investment. Finally, this chapter underscores its role in creating equitable, personalized, and cost-effective learning experiences in modern language education.

Keywords: Machine learning, Adaptive learning, Faculty development, Personalized learning

* * *

1. Introduction

This chapter explores the experiences of Modern Languages and Literatures (MLL) faculty teaching foreign language courses at the University of Central Florida (UCF) as they integrate personalized adaptive learning (PAL) to enhance student learning outcomes and engagement. This chapter provides insights into its implementation and best practices by examining how PAL curricula were designed, expanded, and refined within language education. The goal is to offer real-world insights and practical best practices that can inform educators and administrators seeking to adopt adaptive learning as a transformative technology in language instruction.

Adaptive learning has evolved substantially over the past several decades, with growing application in second language acquisition (SLA). The foundation was laid by B.F. Skinner's (1958) teaching machines, which introduced self-paced, feedback-driven programmed instruction. The 1970s marked the rise of computer-assisted instruction, expanding opportunities for individualized learning through early educational software in language learning (Alderman et al., 1978; Hart, 1995). As technology advanced, adaptive systems began to emerge, using rule-based decision-making to tailor content and practice to learner needs. Over time, these systems grew more sophisticated, incorporating user data, linguistic profiling, and algorithmic logic to better address the complexities of second language learning (Chen & Rebuschat, 2022).

2. Personalized adaptive learning in language education

While most extensively implemented in the fields of science, technology, engineering, and mathematics (STEM), adaptive learning systems are increasingly transforming language acquisition through sophisticated features such as diagnostic profiling, dynamic difficulty adjustment, and individualized feedback (Woo & Choi, 2021; Xie et al., 2019). These tools use speech recognition and real-time data to adjust instruction across domains, including pronunciation, vocabulary, grammar, reading, and writing (Ismail & Wiemer-Hastings, 2023; Tan, 2024; Wilschut et al., 2021a;). Notable examples include MyEnglishLab, which tailors instruction through immediate feedback and adaptive difficulty, and SuccessMaker, which integrates real-time data to personalize lesson sequences (Delgado et al., 2020). Other adaptive language learning applications, such as Duolingo, SlimStampen, and Babbel, use adaptive algorithms to track user progress and adjust learning pathways accordingly (Betaubun et al., 2023; Wilschut et al., 2021a). Intelligent tutoring systems (ITS) have also been developed to mimic human tutors by analyzing student responses and providing targeted interventions (Aleven et al., 2009; Woo & Choi, 2021). These tools collectively demonstrate how technologies are reshaping language education by offering personalized, adaptive, responsive learning experiences.

2.1. Adaptive features

Personalized adaptive learning systems in language education leverage multiple technological components to deliver individualized learning experiences that align with SLA approaches such as input, interaction, and output (Gass & Mackey, 2020). Machine learning (ML) and data analytics are central to this process, integrating three core components: content modeling for linguistic elements, learner modeling for proficiency tracking, and instructional modeling for pedagogical adaptation or scaffolding (Imhof et al., 2020). These models work in concert to analyze learner interactions, progress, and errors continuously, thereby adjusting instructional content dynamically to maintain optimal levels of comprehensible input and scaffold learning. Adaptive systems promptly identify language gaps and patterns essential for acquisition and tailor materials to the learner's current proficiency accordingly.

Natural language processing (NLP) is another key component that supports SLA by enabling adaptive systems to assess pronunciation, grammar, and fluency in real time (Qin & Zhong, 2024). This technology creates meaningful interaction between the learner and the system by evaluating spoken and written responses and providing immediate, personalized corrective feedback. Such feedback prompts learners to attend to their linguistic errors and guides them in refining their output. Techniques like pronunciation assessment, speech-to-text conversion, and semantic analysis facilitate authentic output, giving learners opportunities to produce language and receive targeted, individualized feedback that deepens learning (Litman, 2016).

Adaptive assessments also support SLA by ensuring that learners are challenged at an appropriate level through dynamic adjustment of question complexity. Unlike static assessments, adaptive testing reacts in real-time to learner responses, reinforcing areas of weakness while allowing mastery of familiar content to be acknowledged and bypassed. This approach maintains an optimal level of input that is just beyond the learner's current competence, supporting Krashen (1991)'s comprehensible input hypothesis. Moreover, by providing immediate, personalized feedback, adaptive assessments transform traditional evaluation into an ongoing formative process that guides learning, sustains motivation, and encourages meaningful interaction with the content.

2.2. Benefits and challenges

The literature on adaptive learning highlights several significant benefits in the context of language education. One of the most significant advantages is its impact on student learning outcomes in relation to pronunciation (Shih et al., 2010), vocabulary (Wilschut et al., 2021b), grammar (Baker et al., 2017), reading (Kostolanyova & Simonova, 2020), and writing (Gohar & El-Ghool, 2016). Adaptive learning platforms analyze learners' strengths and weaknesses, customizing instructional materials to address individual needs. Research suggests that adaptive systems enhance language proficiency by providing continuous formative assessment and targeted interventions, leading to greater retention and mastery of language skills (Delgado et al., 2020). Studies have shown that adaptive tools have reduced grading and learning gaps in language learning, especially for minoritized students (Delgado et al., 2020; O'Sullivan & Forgette, 2024). Furthermore, ITS offer real-time feedback and personalized learning pathways, helping learners address their specific language deficiencies more efficiently (Betaubun et al., 2023).

Another key benefit of PAL systems is increased learner engagement. Traditional language learning approaches often struggle to maintain student motivation, particularly in self-paced or online learning environments. Adaptive platforms address this issue by incorporating interactive media, personalized challenges, and real-time feedback mechanisms, which make the learning experience more targeted and dynamic. Features such as algorithmically created content, progress tracking, adaptive difficulty levels, and personalized learning paths help sustain student interest and reduce dropout rates (Dziuban et al., 2018). Research indicates that learners exposed to adaptive language learning tools show higher levels of participation and persistence in their studies as the systems create a more interactive and stimulating learning environment (Delgado et al., 2020; Qin & Zhong, 2024).

Although there is significant enthusiasm for the potential of incorporating adaptive learning in education and growing evidence supporting its effectiveness, its adoption in higher education, particularly within language learning curricula, remains limited (Abadia & Liu, 2021; Mirata et al., 2020). The development of PAL courses requires significant institutional investment, including selecting appropriate platforms, training faculty to integrate adaptive tools into their instruction, and ensuring ongoing technical support (Chen et al., 2025).

Faculty resistance is one significant barrier to the adoption of adaptive learning in language education. Many instructors hesitate to transition from established teaching methodologies to adaptive learning platforms due to unfamiliarity with the technology, skepticism about its effectiveness, and concerns regarding the balance between human interaction and automated instruction (Abadia & Liu, 2021). Language education heavily relies on social interaction, communication practice, and contextual learning, leading to doubts about whether adaptive learning can fully replicate the nuances of instructor-led feedback and peer collaboration. Additionally, the cost associated with implementing adaptive technologies poses a barrier (Betaubun et al., 2023). Licensing adaptive learning platforms can be expensive, and instructors are often reluctant to impose additional financial burdens on students. These challenges underscore the need for institutional support and comprehensive faculty development programs to facilitate the effective integration of adaptive learning technologies in education (Cavanagh et al., 2020).

Despite challenges, adaptive learning has the potential to enhance educational outcomes in language programs by providing personalized instruction that meets the diverse needs of learners. This chapter seeks to provide valuable insights into the implementation process of adaptive learning in language education, highlighting best practices from educators who have effectively redesigned their curricula. The foreign language faculty at UCF embraced this innovative approach to enhance student outcomes in modern language courses.

3. Course redesign initiative

Launched in 2018 by the Center for Distributed Learning (CDL) at UCF, the Course Redesign Initiative (CRI) promoted enhanced student learning experiences by strategically redesigning courses utilizing digital learning methodologies. The timeline for individual CRI projects was one academic year, with three cycles of projects between 2017-2020. This initiative focused on implementing evidence-based instructional practices to improve student outcomes, particularly in challenging courses. The CRI was designed to improve student learning outcomes by (1) increasing course completion rates; (2) improving student retention and overall success; (3) focusing on general education, STEM, and foundational courses; and (4) improving classroom utilization (UCF, n.d.). The goals of the initiative were expansive: to fundamentally transform 100 courses, with 50 of these courses using adaptive learning, ultimately impacting up to 50,000 student enrollments.

In 2020, the CRI was extended for an additional one-year cycle, focusing on group projects (e.g., redesigning sequential courses). The main goals were to (1) enhance successful course completion rates, especially within the General Education Program and STEM courses, and (2) support the persistence of first-time-in-college and transfer students. Additional goals included building up successes and lessons learned, as well as maximizing innovations that emerged during the 2019 coronavirus epidemic. The CRI Extension aimed to promote faculty preparedness and maximize student success by leveraging personalized adaptive learning and program-level redesign. Several of the MLL faculty who participated in this study also participated in the initial CRI.

Participating in the CRI, the MLL faculty redesigned their Elementary Spanish Language and Civilization I and II courses using adaptive learning technology and open educational resources (OER; no cost, shareable educational content). This redesign project aimed to create a more personalized and engaging learning experience for students, many of whom had different levels of prior knowledge in Spanish. The adaptive learning platform, Realizeit, allowed students to progress at their own pace, focusing on areas where they needed more practice and reinforcement. This approach helped students apply their learning in real-world contexts, a key goal for many of them (O'Sullivan et al., 2020). The outcomes of the redesign were significant. In Spanish I, the student success rate increased by 23%, and the withdrawal rate decreased by 7%. Similarly, in Spanish II, the success rate increased by 22%, and the withdrawal rate decreased by 13%. Student satisfaction improved in both courses, with notable increases in end-of-course evaluation scores. These positive results have led to scaling the redesign courses across more Spanish language program instructors and other language courses, such as French at UCF (O'Sullivan et al., 2020). Adaptive learning and OER in these modern language courses improved student learning outcomes and provided a more cost-effective and personalized educational experience. The success of this initiative underscores the potential for adaptive learning technologies to change language education and improve student engagement and achievement (O'Sullivan et al., 2020).

4. The study

Using an explanatory case study design (Yin, 2018), this study explored faculty experiences to understand their reasons for adopting PAL, the instructional challenges they aimed to resolve, and the effectiveness of adaptive learning strategies. By exploring faculty perspectives and implementation practices, the results of this study offer insights into how adaptive learning can reshape language instruction and inform best practices for educators, administrators, and policymakers.

4.1. Context and participants

The state of Florida currently requires college students to complete one year of foreign language study or its equivalent, prompting many to enroll in foreign language courses (Florida Department of Education, 2025). With enrollment consistently above 65,000 students, UCF's Department of Modern Language and Literatures offers courses in Arabic, Chinese, Russian, French, German, Haitian Creole, Italian, and Japanese, with bachelor's degrees offered in

Spanish, French, and Latin American, Caribbean, and Latinx Studies and graduate degrees offered in Spanish and Teaching English to Speakers of Other Languages. Because the university offers many MLL courses online, most MLL faculty have received training in online course development and delivery. The MLL faculty have demonstrated an ongoing commitment to growing their skills and knowledge around the tools and strategies associated with online course development and teaching.

The participants of this study consisted of nine MLL faculty who met the following criteria: (1) they had completed the PAL training offered through CDL, and (2) they had designed, developed, and delivered at least one PAL module for use in an MLL course. Of these nine faculty members, eight taught Spanish, with one instructor teaching Portuguese and the other teaching French. One faculty member taught Portuguese and no other language. Most of the participants (6) reported teaching their native language; three reported teaching a language that was not their native language. Nearly all of the faculty participants in this study had extensive experience as educators; several reported 30 or more years, while others reported 20 or more. Two participants reported 10 years or less of experience.

4.2. Procedures and data analysis

This project aimed to identify the qualities of PAL that inspire MLL faculty to invest significant time and resources into developing and delivering PAL instruction for their students. Because we were interested in exploring the motivations and experiences of our study population, we used an explanatory case study design (Yin, 2018). The benefits of using a case study design include using follow-up questions to explore individual participants' unique experiences and perspectives.

The first step towards our goal involved developing the study protocol, consisting of the questions posed to the study's participants. The protocol focused on the participants' motivation for adopting PAL, their experiences using PAL-oriented strategies, and the outcomes of using PAL in their courses. In addition to the demographic questions, the protocol included the following three questions: (1) What motivated you, a member of the modern languages faculty, to use PAL? (2) What problem or problems were you attempting to solve by incorporating PAL into your course or courses? And (3) What features of PAL do you think are especially well-suited to L2 education?

At the same time, we were developing our study protocol, we began recruiting MLL faculty who had developed or used PAL in at least one course. A total of 13 MLL faculty qualified to participate in our study; as previously stated, nine MLL faculty accepted our invitation to participate in the study. We conducted all interviews using Zoom, providing a complete transcript of each conversation.

After completing all interviews, we assigned topic-based codes to the individual transcripts and combined interview comments according to shared codes.

5. Results

Faculty collaboration, institutional initiatives, and the need to address key challenges in student learning have driven the adoption of innovative teaching tools in language courses. Faculty members were inspired by their colleagues' successful implementation of Realizeit and the CRI, motivating them to explore adaptive learning technologies like the PAL program. A primary challenge in Spanish language instruction has been the varying proficiency levels among students, leading to inefficiencies in traditional classroom settings. The PAL program emerged as a solution, allowing for personalized learning that adapts to individual student needs while reducing redundant coursework. Additionally, the high cost of textbooks posed a financial burden, prompting faculty to seek more affordable, accessible learning materials. By integrating PAL with affordability initiatives, the department has created a more inclusive and effective learning environment where students can progress at their own pace, engage in continuous practice, and receive targeted support to improve their language proficiency.

5.1. Collaboration with colleagues

Several participants identified the role of collaboration and peer influence in adopting new teaching methods and tools. They were motivated by colleagues who had successfully integrated Realizeit into their classes. Initially, the tool was used by MLL faculty as an add-on to Canvas (the university's learning management system) for review purposes. Still, after positive experiences, they decided to implement it fully. The CRI similarly inspired MLL faculty to explore Realizeit. One participant reported that "a lot of my colleagues had done the [Course] Redesign Initiative, and I know in Spanish they seem to have been successful. So, we were thinking for French that maybe we could do the same thing." Another faculty member from the Spanish department was similarly inspired by the CRI, noting that "we applied and got it [CRI], and that was a big motivating factor, interacting and collaborating with a colleague." A participant also revealed being motivated by department-wide interest as other colleagues explored using Realizeit in different language courses.

5.2. Bridging the gap between skill level and course level

The primary challenge in Spanish language courses is the varying proficiency levels of students, even within the same course, such as SPN3300 (Advanced Spanish Grammar and Composition) or introductory courses like Spanish 1120 and 1121. Some students already possess knowledge of specific topics, while others struggle with foundational concepts, leading to frustration and inefficiencies in learning. One of the study participants observed that "a student may be placed in SPN1120, but they still know some stuff. So, why should they be punished by having to do 100 activities about something they already know?" To address this problem of various levels of language competency, instructors initially attempted to create differentiated activities for various proficiency levels in face-to-face classrooms, but this proved time-consuming:

> And one of the things that we were trying to incorporate in a classroom face to face - the dividing the class in different levels and it takes a lot of time to do that. It takes a lot of time to prepare like certain activities that's good at one level and then preparing other activities different for students that work at another level, more and more advanced level. And then, when they offer us this opportunity (PAL), we thought that this was going to solve the issue, and, in a certain way, it has.

Implementing the PAL program was seen as a potential solution, allowing for personalized learning that adapts to students' existing knowledge, making them more successful while minimizing repetitive tasks.

Another issue is the lack of a departmental placement exam, which makes it difficult to place students in courses that match their skill level accurately. Placement errors can put students in classes that are either too advanced or not challenging enough, further complicating the learning process. Misplacement can place students in classes that are either too advanced or not challenging enough, further complicating the learning process. PAL aimed to bridge these gaps by providing remedial activities and reviewing first-year material to ensure that students were better prepared for intermediate and advanced courses.

Similarly, a participant observed that for students who may be between course levels "or have some exposure to the language, but not enough to go on to another level, using a program like this would allow for them to be more successful in the class, and not have to repeat things that they already know." Overall, the goal is to create a more efficient and supportive learning environment where students can progress at their own pace, focusing on areas needing improvement without being penalized by redundant coursework.

5.3. Opportunities to practice

The PAL program was implemented as a tool to enhance student learning by allowing for unlimited practice, self-paced progress, and continuous revision.

One of its key benefits is that students can repeatedly engage with grammar exercises, revisiting concepts they struggle with until they fully understand them. This flexibility helps reduce frustration and makes learning more effective than traditional worksheets or static materials. "I was motivated to use PAL because it seemed to me the vehicle that allowed students to practice as much as they needed to, to learn that piece of information regarding the grammar, and so on." PAL offers an interactive and adaptive approach to grammar practice, making it more engaging than rote memorization. Students are encouraged to drill grammatical rules and structures like how one would practice math formulas. They understand that they can go back and continue practicing the concepts they don't fully understand. The ability to revise scores and redo activities without penalty further motivates students to improve their understanding, as their grades can only improve with additional effort. Study participants shared their enthusiasm for improved opportunities to practice new skills, noting that "language has rules the same way math does, and so you have to practice the rules... and drill and drill and drill." Another participant observed that "students have to do a lot of grammar practice to master the [necessary] skills. So, PAL really helps them to practice, and then go back and practice more if they like."

Instructors also appreciate that PAL keeps practice materials open throughout the semester, allowing students to return to them even after due dates. This approach supports long-term mastery rather than a short-term, topic-by-topic learning model. Additionally, PAL functions as a workbook, a review space, and a practice tool all in one, reinforcing the idea that language learning requires continuous engagement rather than one-time exposure. Ultimately, PAL provides a structured yet flexible environment that aligns with the repetitive nature of mastering a second language, helping students build confidence and proficiency at their own pace.

5.4. Cost of materials

A significant issue identified in Spanish language courses was the high cost of required materials. Students spent an average of $300 on publisher packages for courses like Spanish 1120 and 1121, even if they only needed materials for one of the courses. This expense was particularly burdensome for students starting at the intermediate level, who still had to purchase expensive digital platforms and textbooks despite potentially only using them for a short period:

Obviously, I understand that textbooks are expensive. Textbooks were expensive when I was in school, but to see data showing how those rates have increased, and then, knowing that I had students who are waiting on financial aid to get the book, and then also knowing you know how I had to work with those students to try to keep them in the course without having the textbook, and how they struggled.

To address this, faculty aimed to develop course materials that mimicked publisher content but were freely accessible, eliminating the need for costly textbooks. The goal was not only to reduce financial strain on students but also to provide an alternative teaching method that moved away from traditional textbooks.

Affordability became a central focus after the department joined the CRI, aligning with institutional goals to lower student expenses. Data tracking on student enrollment and spending showed the increasing costs of textbooks and the struggles students faced when waiting for financial aid to afford them. Many students expressed gratitude for cost-saving measures, as the new approach helped them stay in the course without financial barriers. The department successfully reduced textbook costs under the $60 threshold through these efforts, qualifying for the affordability initiative. This integration of PAL with affordability considerations was a significant success, benefiting both students and instructors.

5.5. Improving engagement and outcomes

Participants emphasized opportunities to enhance the online learning experience, particularly in elementary language courses, by improving student engagement. They highlighted the immediate instructional feedback provided by Realizeit, particularly on machine-gradable assignments, which helps students identify areas for improvement and enhances their learning process.

> Realizeit is especially well-suited for a language as a second language and education. Oh, I think just the immediate feedback that the students receive on the machine-gradable assignment. I think that's really helpful for them because it shows them right away where they need to practice more.

The platform also provides clear expectations for students, reducing confusion and making learning objectives more transparent. As a result, students likely feel more comfortable and find the course structure more predictable, leading to a more cohesive learning experience. Also, the structured approach has contributed to a more organized course, with students recognizing and appreciating this improvement. A participant noted that they received student feedback indicating that "the [Realizeit] course was really well organized. And so, for me, that was a huge success." The opportunity to work independently on a computer with immediate feedback further enhanced student motivation and engagement, making online learning more effective and accessible.

A common goal that connected all of our study participants was the improvement of student outcomes, as demonstrated by lower rates of course drop, withdrawal, and failure. Participants observed a decrease in drop and withdrawal rates along with a drop in failure rates, indicating that more students were staying engaged and completing their courses. Faculty participants also observed how PAL worked with the instructor's content to improve student success:

> I believe the combination of having Realizeit as a review with having content in the courses, or me delivering the content in the classroom setting. I think the combination of all this has given a great experience in the language, and from my perspective it has been successful.

One of the more experienced participants noted that "withdrawals decreased the first semester that I taught with PAL; my student [performance reviews] were the best I have ever had, and, at that point, I had been already teaching at UCF for like 23 years." Table 8.1 summarizes the participants' motivations to implement PAL.

Motivation	Key Findings
Collaboration & Peer Influence	Faculty were influenced by colleagues' success with Realizeit and the Course Redesign Initiative (CRI). Shared experiences encouraged broader adoption.
Bridging Skill & Course Level Gaps	Students in the same course had varied proficiency levels. PAL allowed personalized learning, minimizing redundant activities and enhancing efficiency.
Departmental Placement Exam Issues	Lack of proper placement exams led to students being in courses that were too easy or too challenging. PAL provided remediation and reinforcement.
Opportunities for Practice	PAL enabled unlimited, self-paced practice, reinforcing learning and reducing frustration compared to static materials.
Long-Term Learning Access	PAL materials remained accessible throughout the semester, supporting mastery beyond one-time exposure.
High Cost of Course Materials	Students spent around $300 on textbooks and digital resources. Faculty sought affordable alternatives, including Open Educational Resources (OER).

Affordability Initiatives	CRI aligned with institutional goals to reduce student expenses, leading to textbook costs dropping below $60.
Improved Engagement & Feedback	Realizeit provided immediate instructional feedback, making learning more transparent and engaging.
Student Satisfaction & Success	Faculty observed reduced withdrawal and failure rates with PAL. Students appreciated structured course design and independent learning opportunities.

Table 8.1. Participants' motivations to implement PAL in their courses

6. Implementing personalized adoptive learning: faculty challenges

Despite their enthusiasm for PAL, study participants reported significant challenges in its implementation. At first, they were eager to move away from traditional textbooks and create their own materials, but the process proved far more time-consuming than expected. The complexity of the Realizeit platform became a significant hurdle. Realizeit required substantial support from instructional designers for tasks as simple as editing quiz questions. Some participants felt frustrated by their reliance on external support and wished for a more user-friendly version designed for everyday instructors. As reported by one of the study participants:

> In my opinion, it's [Realizeit] not super user-friendly. In Realizeit, I needed a lot of help from my instructional design(er). At one point, I knew I made a mistake in one of the questions, and I was not even able to fix it. I had to contact my instructional designer and ask for help – that's the part that I don't like.

To help mitigate these challenges, PAL instructional designers worked intensively alongside faculty throughout the course design and development process. Faculty members were required to complete a dedicated training course focused on effective PAL course design and delivery. This collaborative approach aimed to ensure faculty felt supported while learning to navigate the platform's features, develop effective adaptive content, and ultimately deliver a high-quality personalized learning experience to their students. By fostering close partnerships and building faculty capacity, the instructional design team sought to overcome technical hurdles and empower instructors to leverage adaptive learning tools more confidently and independently in their teaching.

Another challenge was navigating within PAL modules. Students could retake activities but often struggled to find lesson content, leading some to complete exercises without engaging with the instructional material. Faculty also found the feedback system lacking; students received basic right-or-wrong responses without direct links to relevant lessons, making it difficult for them to review and reinforce their learning. At least one participant said, quite simply, that "it [Realizeit] is not an easy program to use at all."

In addition to navigating challenges, Realizeit lacks essential features for foreign language learning. Unlike the university's learning management system (LMS; Canvas), which allows for voice recordings to assess pronunciation, Realizeit lacks a built-in voice recognition tool. Since pronunciation is a crucial component of language acquisition, participants found this limitation particularly problematic. One participant explained:

> In foreign languages, we have to assess all the skills, and one of the features that is in [Canvas] that we don't have in PAL. It's like the recording, so we can assess pronunciation, and that's something that I would say we needed in there. We need a voice recognition because part of the problems is there is nothing in there for pronunciation.

The difficulties also extended to assessment management; locating and correcting quiz questions within Realizeit's extensive word banks was cumbersome and inefficient.

Cost was another pressing concern for this study's participants. Faculty expressed their commitment to using free, online textbooks whenever possible, allowing students to complete coursework without financial burden. However, if the cost of Realizeit shifted from the university to student users, this change would be unacceptable for most participants. While institutional funding covered PAL courses for the current semester, uncertainty about future funding left participants questioning its long-term viability.

7. Conclusion

The integration of PAL in foreign language courses at UCF has demonstrated significant potential in enhancing student outcomes, fostering engagement, and addressing diverse learner needs. The findings of this study underscore the significant benefits of PAL in language education, aligning closely with the insights presented in the literature review. First, adaptive learning technology has been shown to effectively target students at specific language proficiency levels and address learning gaps. By providing personalized instruction and ample opportunities for practice, students can progress at their own pace, reinforcing their skills through continuous feedback and adaptive assessments. This individualized approach enhances language acquisition by ensuring learners receive the appropriate level of challenge and support based on their needs.

Strong institutional support is another key factor contributing to PAL's successful adoption and expansion in language education. At UCF, internal grant opportunities have been crucial in fostering faculty collaboration and innovation. These grants provide language instructors with the necessary resources to explore, develop, and refine adaptive learning strategies, creating a sustainable model for long-term implementation. The availability of institutional funding and structured faculty development initiatives ensures that adaptive learning is not only introduced but continuously improved and expanded.

Based on the findings of this study, the authors suggest the following best practices for adaptive learning implementation in language education and other disciplines (Table 8.2):

Recommendations	Description of Best Practices
Consider open educational resources (OER)	To reduce costs, institutions should prioritize adaptive platforms that allow for OER integration, making learning more accessible and affordable for students. Provide faculty with curated OER repositories and training on selecting or adapting open content for adaptive pathways.
Promote faculty collaboration	Encourage collaboration between experienced PAL faculty members and faculty new to PAL. Send regular email reminders with "tips and tricks," success stories, and resources, and help connect faculty for sharing adaptive learning designs and solutions.
Provide structured faculty training	Offer dedicated tiered professional development programs and collaborative opportunities for faculty to explore best practices in adaptive learning and instructional design and reduce learning curve. Provide one-on-one coaching on adaptive pathways and assessment design.
Manage costs to maintain student access	Align PAL with institutional affordability initiatives to minimize financial barriers for students by evaluating platform licensing models and encouraging OER use.
Monitor and evaluate effectiveness	Assess adaptive learning implementation at regular intervals to measure student success, identify areas for improvement, and refine instructional strategies. Include formative check-ins with faculty and student feedback surveys to guide adaptive design refinements.
Leverage institutional grants and support networks	Encourage faculty collaboration through institutional grants and interdisciplinary partnerships that foster the development of innovative adaptive learning initiatives. Foster interdisciplinary partnerships and create faculty learning communities to sustain innovation and share effective models.

Table 8.2. Best practices in personalized adaptive learning implementation

As artificial intelligence (AI) technologies continue to advance, integrating adaptive learning with other AI-driven tools will further enhance language education by providing more personalized, responsive, and efficient learning experiences. The ability of these technologies to dynamically adjust instruction based on individual student progress will play a crucial role in improving engagement, bridging learning gaps, and fostering greater autonomy in language acquisition. However, the successful implementation of PAL requires careful planning, sustained institutional investment, and a strong commitment to faculty collaboration. By prioritizing these elements, adaptive learning can become a transformative force in language education, equipping students with the necessary skills while making high-quality learning experiences more accessible, equitable, and cost-effective.

References

Abadia, R., & Liu, S. (2021). Low adoption of adaptive learning systems in higher education and how can it be increased in fully online courses. In M. M. T. Rodrigo, S. Iyer, & A. Mitrovic, *Proceedings of the 29th International Conference on Computers in Education, Asia-Pacific Society for Computers in Education*, pp. 569-578. Asia-Pacific Society for Computers in Education. https://find.library.unisa.edu.au/discovery/fulldisplay/alma9916619141101831/61USOUTHAUS_INST:ROR

Alderman, D. L., Appel, L. R., & Murphy, R. T. (1978). PLATO and TICCIT: An evaluation of CAI in the community college. *Educational Technology, 18*(4), 40–45. https://www.jstor.org/stable/44419177

Aleven, V., Mclaren, B. M., Sewall, J., & Koedinger, K. R. (2009). A new paradigm for intelligent tutoring systems: Example-tracing tutors. *International Journal of Artificial Intelligence in Education, 19*(2), 105–154.

Baker, R., Wang, F., Ma, W., & Zheng, S. (2017). *Effectiveness of an online language learning platform in China.* https://www.semanticscholar.org/paper/Running-head-%3A-EFFECTIVENESS-OF-AN-ONLINE-LANGUAGE-Baker-Wang/69db7720f23e73e3148a1fef44968777fa99b45f

Betaubun, M., Rokhmah, D. E. L., & Budiasto, J. (2023). Personalized pathways to proficiency: Exploring the synergy of adaptive learning and artificial intelligence in English language learning. *Technium: Romanian Journal of Applied Sciences and Technology, 17,* 60–66. https://doi.org/10.47577/technium.v17i.10047

Cavanagh, T., Chen, B., Lahcen, R. A. M., & Paradiso, J. (2020). Constructing a design framework and pedagogical approach for adaptive learning in higher education: A practitioner's perspective. *The International Review of Research in Open and Distributed Learning, 21*(1), 173–197. https://doi.org/10.19173/irrodl.v21i1.4557

Chen, B., Tojo-Raible, J., Lloyd, J., & McNulty, R. (2025). Designing adaptive learning for higher education: A comprehensive ADDIE approach. In *The Sage Handbook of Higher Education Instructional Design.* Sage.

Chen, X., & Rebuschat, P. (2022). ICALL offering individually adaptive input: Effects of complex input on L2 development. In *Proceedings of the International Conference on Artificial Intelligence in Education.*

Delgado, H. O. K., Fay, A. de A., Sebastiany, M. J., & Silva, A. D. C. (2020). Artificial intelligence adaptive learning tools: The teaching of English in focus. *BELT - Brazilian English Language Teaching Journal, 11*(2), Article 2. https://doi.org/10.15448/2178-3640.2020.2.38749

Dziuban, C., Moskal, P., Parker, L., Campbell, M., Howlin, C., & Johnson, C. (2018). Adaptive learning: A stabilizing influence across disciplines and universities. *Online Learning, 22*(3), 7–39. https://doi.org/10.24059/olj.v22i3.1465

Florida Department of Education. (2025). *World languages (foreign languages).* Retrieved February 1, 2025, from https://www.fldoe.org/academics/eng-language-learners/world-languages-foreign-languages.stml

Gass, S. M., & Mackey, A. (2020). Input, interaction, and output in L2 acquisition. In *Theories in Second Language Acquisition* (3rd ed.). Routledge.

Gohar, R. H., & El-Ghool, R. M. (2016). Designing an adaptive learning environment to improve writing skills and usability for EFL students at the faculty of education. *International Journal of Internet Education, 15*(1), 63–93. https://doi.org/10.21608/ijie.2016.3682

Hart, R. S. (1995). The Illinois PLATO Foreign Languages Project. *CALICO Journal,* 12(4), 15–37. https://www.jstor.org/stable/24147585

Imhof, C., Bergamin, P., & McGarrity, S. (2020). Implementation of adaptive learning systems: Current state and potential. In P. Isaias, D. G. Sampson, & D. Ifenthaler (Eds.), *Online Teaching and Learning in Higher Education* (pp. 93-115). Springer Nature Switzerland. https://doi.org/10.1007/978-3-030-48190-2_6

Ismail, D., & Wiemer-Hastings, P. (2023). Emotionally adaptive intelligent tutoring system to reduce foreign language anxiety. In *Proceedings of the International Conference on Artificial Intelligence in Education.*

Kostolanyova, K., & Simonova, I. (2020). Adaptive e-learning model for learning English as a second/foreign language. *International Journal of Innovation and Learning, 27*(3), 241–257.

Krashen, S. (1991). The input hypothesis: An update. In J. Alatis (Ed.), *Linguistics and Language Pedagogy: The State of the Art* (pp. 409–431). Georgetown University.

Litman, D. (2016). Natural language processing for enhancing teaching and learning. *Proceedings of the AAAI Conference on Artificial Intelligence, 30*(1), 4170-4176. https://doi.org/10.1609/aaai.v30i1.9879

Mirata, V., Hirt, F. S., Bergamin, P., & Westhuizen, C. P. V. D. (2020). Challenges and contexts in establishing adaptive learning in higher education: Findings from a Delphi study. *International Journal of Educational Technology in Higher Education, 17,* Article 2. https://doi.org/10.1186/s41239-020-00209-y

O'Sullivan, P., & Forgette, C. (2024). *The impact of digital learning on minoritized and poverty-affected college students: 12 instructors' stories.* Every Learner Everywhere. https://www.everylearnereverywhere.org/resources/the-impact-of-digital-learning-on-minoritized-and-poverty-affected-college-students-case-studies/

O'Sullivan, P., Voegele, J., Buchan, T., Dottin, R., Goin Kono, K., Hamideh, M., Howard, W. S., Todd, J., Tyson, L., Kruse, S., de Gruyter, J., & Berg, K. (2020). Adaptive courseware implementation: Investigating alignment, course redesign, and the student experience. *Current Issues in Emerging eLearning, 7*(1), 31-34. https://scholarworks.umb.edu/ciee/vol7/iss1/6

Qin, L., & Zhong, W. (2024). Adaptive system of English-speaking learning based on artificial intelligence. *Journal of Electrical Systems, 20*(6s), 267–275. https://doi.org/10.52783/jes.2637

Shih, C., Lu, H.-Y. D., Sun, L., Huang, J.-T., & Packard, J. (2010). An adaptive training program for tone acquisition. *Speech Prosody 2010,* paper 981. https://doi.org/10.21437/SpeechProsody.2010-72

Skinner, B. F. (1958). *Teaching machines. Science, 128*(3330), 969–977. https://doi.org/10.1126/science.128.3330.969

Tan, J. (2024). An empirical study of adaptive learning algorithm based on intrinsic motivation in English online teaching and learning. *Journal of Electrical Systems.* https://doi.org/10.52783/jes.3088

University of Central Florida. (n.d.). *DL CRI: Original 3-year initiative.* Retrieved February 2, 2025, from https://digitallearning.ucf.edu/ilab/cri/original-dlcri/

Wilschut, T., Sense, F., van der Velde, M., Fountas, Z., Maaß, S. C., & van Rijn, H. (2021). Benefits of adaptive learning transfer from typing-based learning to speech-based learning. *Frontiers in Artificial Intelligence, 4.* https://doi.org/10.3389/frai.2021.780131

Wilschut, T., Sense, F., & van Rijn, H. (2021). Speaking to remember: Model-based adaptive vocabulary learning using automatic speech recognition. *Computer Speech & Language, 70,* 101237. https://doi.org/10.1016/j.csl.2023.101578

Woo, J.-H., & Choi, H. (2021). Systematic Review for AI-based Language Learning Tools. *Journal of Digital Contents Society, 22*(11), 1783–1793. https://doi.org/10.9728/dcs.2021.22.11.1783

Xie, H., Chu, H. C., Hwang, G. J., & Wang, C. C. (2019). Trends and development in technology-enhanced adaptive/personalized learning: A systematic review of journal publications from 2007 to 2017. *Computers & Education, 140,* 103599. https://doi.org/10.1016/j.compedu.2019.103599

Yin, R. K. (2018). *Case study research and applications: Design and methods.* Sage.

Zylich, B., & Lan, A. S. (2021). Linguistic skill modeling for second language acquisition. In *Proceedings of the International Conference on Learning Analytics and Knowledge.* https://people.umass.edu/~andrewlan/papers/21lak-slam.pdf

Part III.
Pedagogical strategies and reflections for language educators

Chapter 9

Best practices and challenges in the use of VR in the language classroom

Samanta Buffa
The Ohio State University

Kevin Richards
The Ohio State University

Rebecca Bias
The Ohio State University

Abstract

This chapter discusses best practices, challenges, and insights from a virtual reality (VR) project using Meta Quest 2 headsets in an intermediate Italian language course at The Ohio State University. Based on instructed second language acquisition (ISLA) and the concept of task from the task-based language teaching (TBLT) approach, VR was used to develop activities where students could explore and manipulate objects. The chapter reviews the steps to create and implement a project, including development, technical issues, classroom experiences, and lessons learned. We recommend practices to address economic, technical, and personal limitations, such as cybersickness, student-instructor reactions, and time management.

Keywords: Virtual reality, Language teaching and learning, Teaching with technologies, Instructed second language acquisition, Task-based language teaching

* * *

1. Introduction

Due to the emerging research and enthusiasm surrounding virtuality reality (VR) headsets for professional and academic applications, there is increased interest in identifying and establishing best practices for their implementation in the classroom. In addition, the introduction of budget-friendly headsets has made VR a more mainstream and viable technology for language learning. While learning platforms appeared to meet a growing demand, they often did not match instructor-specific learning goals, prompting us to develop targeted content. In this chapter, we draw upon contemporary research, as well as insights gained from original research conducted in the fall of 2020, to provide an orientation to VR development for language acquisition.

Recent literature has provided support for the implementation of VR in language learning. For example, studies on the application of VR in science, technology, engineering, and mathematics (STEM) have shown improvements in learners' mood, memory, understanding, and motivation (Allcoat & von Mühlenen, 2018; Graeske & Sjöberg, 2021; Johnson-Glenberg et al., 2021; Tai et al., 2020). "In addition… VR and AR can be used among students with zero previous knowledge up to an intermediate and advanced level of fluency, with any L2 language" (Pataquiva & Kilmová, 2022, p. 43). Based on these encouraging results, we conducted a pilot study with twenty undergraduate students at the American Council on the Teaching of Foreign Languages (ACTFL) Intermediate Level at The Ohio State University. We created virtual worlds in VRChat Creator Companion for Unity, focusing on key Italian grammar and vocabulary topics.]

The development of this project led to the central thesis of this chapter, which is not only to discover best practices for using VR in the language classroom, but also to highlight challenges and identify solutions for other instructors interested in developing applications. Salient factors include economic, technological, and temporal limitations, as well as cybersickness and personnel constraints. This chapter progresses through three focal areas: (1) pedagogical rationale for VR in second language acquisition (SLA), (2) evidence-based best practices, and (4) common challenges with recommendations. We begin with a discussion of the pedagogical foundations of this study, which is essential, particularly given the hype around VR and the opportunities in gamification.

2. Pedagogical rationale for VR in SLA

The importance of applied SLA theoretical frameworks remains at the forefront when implementing any innovative technology in the classroom. Among the most promising of these technologies is VR, which offers learners opportunities to interact in visually rich, spatially embodied, and communicatively meaningful environments. In the context of SLA, VR does not stand apart from theory-driven instruction; on the contrary, it provides an effective medium through which to operationalize foundational theories such as the interactionist approach, sociocultural theory, and task-based language teaching (TBLT). This section outlines core principles of these theories and demonstrates how virtual learning environments, particularly those facilitated through VR, support key SLA processes, including noticing, comprehensible input, and anxiety reduction.

2.1. The interactionist approach

VR is well-suited to support interactional processes central to the Interactionist Approach to SLA, as it allows for embodied, sensory-rich experiences. This approach asserts that learning occurs through interaction, particularly when learners negotiate meaning, receive feedback, and have opportunities to modify output. Long's (1996) *Interaction Hypothesis* emphasizes that such interactions help acquisition by providing comprehensible input, eliciting learner output, and prompting learners to notice gaps in their interlanguage.

In virtual environments (VEs), learners use avatars, peers, or chatbots—alongside gestures, spatial relationships, and visual props—to scaffold understanding. Tai et al. (2020) compared Taiwanese English as a foreign language (EFL) students who either navigated Mondly's VR restaurant scenario or watched equivalent videos. Those in VR outperformed the video group on both immediate and delayed vocabulary posttests and reported higher motivation and better lexical comprehension due to immersive contextual cues.

VEs encourage learners to stay engaged and respond authentically, which sustains interaction and deepens processing (Özgün & Sadik, 2023). Providing feedback, either through system-generated prompts or peer collaboration, is another key mechanism in the Interactionist Approach. In these tasks, feedback can take various forms, such as visual or auditory indicators of success (e.g., correct object placement, sound, or recorded message) or support from other users. These moments mirror the feedback loops that Gass and Mackey (2007) describe, where communicative breakdown and repair trigger the noticing of form. Furthermore, such environments support modified output by giving learners opportunities to reformulate their utterances in response to feedback or interactional demands. In doing so, they are encouraged to push their language beyond existing competence, a process shown to promote acquisition (Swain, 1985).

2.2. Sociocultural theory and mediation in immersive contexts

Sociocultural theory, rooted in the work of Vygotsky (1978), posits that social negotiation is at the heart of learning. Language development occurs as learners work together, often with guidance from a more capable other, within their zone of proximal development (ZPD). VR platforms that support multi-user collaboration and instructor intervention offer rich opportunities for scaffolded learning, dynamic assessment, and co-construction of meaning. In such VR tasks, instructors can pair learners with peers to promote support and collaboration. This scaffolding can take the form of linguistic modeling, hints, or task assistance, all of which mediate learner performance and promote internalization of new forms. For example, in a VR escape room, a more advanced peer might model a phrase like "Where are you?" and help a partner repeat it when it is their turn to be the guide. Over time, instructors gradually withdraw external support as learners become more autonomous.

Some VR environments also lend themselves to mediated, metalinguistic reflection. For example, on the ImmerseMe platform, learners can pause, replay interactions, or reflect on feedback received during simulations (Uebiyev & Facer, 2025). This metacognitive engagement supports deeper processing and aligns with sociocultural principles of internalization through guided participation. After completing the VR experience, students have time to reflect on the role the technology played in their learning; this approach avoids relegating it to a mere tool, instead recognizing it as an integral part of instruction (Gonzáles-Lloret & Ortega, 2014; Smith & Gonzáles-Lloret, 2021).

Finally, sociocultural theory emphasizes the emotional dimension of learning. VR's immersive design contributes to lowering learners' affective filter by making interactions more playful, purposeful, and embodied. Dooly et al. (2023) conducted a pilot study focused on identifying the kind of language produced by students in face-to-face and VR interactions. They found the language level of interaction in VR to be higher than expected (B1-level students performing at A1 or A2), even though it was less accurate when compared to in-person oral exchanges. Overall, the results showed an improvement in oral proficiency and increased curiosity through the use of VR. By situating learners in meaningful, engaging, and supportive contexts for language use (Dede, 2009), VR brings these theoretical principles to life in ways that might be more difficult to achieve in traditional classrooms.

2.3. Task-based language teaching

TBLT is an approach to communicative language teaching originating in the mid-1980s to help students acquire communicative abilities. In broad terms, it uses authentic, real-world tasks that are meaningful to learners' future goals, with a focus on meaning—not strictly grammar or vocabulary acquisition (Ahmadian & Long, 2021; Ellis et al., 2019; Jackson, 2022; Long, 2015; Nunan, 2015). TBLT is rooted in the belief that language learning is most effective when learners engage in meaningful communication and tasks that reflect authentic language-use situations. In practice, students learn by doing through hands-on experiences that connect to real-world tasks that can be reproduced. Such experiences will "increase the likelihood that abilities learned in the classroom will transfer to the world outside [...] because what is learned is context embedded, better integrated into long-term memory, and more easily recalled" (Long, 2015, p. 307).

Although scholars provide different definitions of *task* (Ellis et al., 2019; Jackson, 2022; Long, 2015; Nunan, 2015), they share several common characteristics. Tasks should be authentic, relevant, and engaging, promoting effective communication and problem-solving. Smith and McCurrach (2021) emphasize that VR provides immersive, multisensory experiences, making it an optimal medium for task-based language instruction and experiential learning. These findings reinforce the TBLT notion that tasks grounded in meaningful virtual interaction promote both form-focused and fluency-based learning. One of Long's (2015) methodological principles—promoting learning by doing—seems especially fitting in virtual worlds. Long (2015) explicitly refers to learners personally experiencing tasks "through the eyes and the hand to the brain" (p. 307). If this is true in the face-to-face (F2F) language classroom, it is even more applicable in VR. Recreating a context, utilizing the immersive and kinesthetic affordances of virtual spaces, and providing total immersion can enhance the integration of knowledge into long-term memory, thereby facilitating recall.

Smith and McCurrach (2021) advocate for the effectiveness of VR in facilitating TBLT, as it "holds the potential to provide incidental learning experiences by facilitating communicative, socio-physical interactions across a host of language learning domains" (p. 153). They assessed three upper-intermediate EFL students playing the VR game "*Keep Talking and Nobody Explodes*," where one student in VR defused a bomb using instructions from two peers in the target language, and the group completed the task. In the debrief, students noted that the VR task required precise communication but not explicit focus on form, supporting TBLT's principles of incidental learning. Drawing on Smith and McCurrach (2021), we argue that VR's immersive, collaborative environments make it an effective medium for task-based instruction, fostering subconscious language acquisition and experiential learning.

2.4. VR affordances aligned with SLA outcomes

The pedagogical potential of VR in SLA is increasingly supported by empirical research demonstrating alignment with key learning outcomes such as noticing, comprehensible input, and reduced language anxiety. This section explores each outcome through theoretical framing and classroom-based examples.

Schmidt's (1990) noticing hypothesis contends that learners must attend to linguistic features in the input to acquire them. VR environments, with their multimodal richness and interactivity, can promote such noticing more effectively than traditional instructional tools. In a study by Lin and Wang (2021), Taiwanese university students learning English engaged in a VR vocabulary training module where they completed tasks such as arranging items in a virtual room and following spoken instructions. The combination of auditory input with visual and spatial cues helped students notice and retain new lexical items more effectively than a non-VR control group. Learners reported that the environment pushed them to process word–object relationships because of the physical and task-based interaction with items.

Similarly, in a classroom implementation by Lan (2020), students learning Mandarin used a VR system that simulated a restaurant scenario. Learners had to navigate a menu, place an order, and respond to culturally appropriate small talk. Repeated exposure to key vocabulary and sentence structures within the same context, along with immediate corrective feedback, promoted noticing without interrupting task flow. For example, when a student used incorrect word order, the system rephrased the input using the correct syntax, allowing the student to notice the target form. These studies suggest that noticing is enhanced when linguistic forms are embedded in task-driven interaction rather than isolated or explicitly taught. VR's ability to provide repeated, meaningful exposure to forms in context, paired with built-in feedback, supports this process in ways traditional input cannot easily duplicate.

In general, students not only need to notice, but also require exposure to comprehensible input. Krashen (1985) highlighted that this input should be slightly beyond the learner's current competence (i+1) and made understandable through context, which is crucial for acquisition. VR excels at delivering this kind of input by offering rich, multimodal contexts in which language is directly tied to visual, spatial, and interactive elements.

VR environments can allow for adjustable difficulty, enabling instructors to scaffold input for learners at different proficiency levels. Instructors can choose scenarios that gradually increase linguistic complexity or offer optional support features such as subtitles, slower speech, or translation glosses. Such adaptations help ensure that learners are consistently challenged, but not overwhelmed—aligning with Vygotsky's (1978) ZPD.

Another barrier students often face is learner anxiety, which hinders performance and persistence in language learning (Horwitz, 1988, 2001; Horwitz et al., 1986). As mentioned earlier, similar results were achieved by Dooly et al. (2023). The repeatable nature of VR scenarios allows learners to build fluency and confidence before engaging in live communication. As affective filters are lowered, learners are more receptive to input and more likely to engage with the target language.

The empirical studies from diverse language-learning contexts presented so far confirm that VR environments support SLA outcomes by leveraging affordances in ways traditional classrooms cannot easily replicate. VR promotes noticing by embedding language in meaningful, interactive tasks that highlight linguistic forms. It enhances comprehensible input through multimodal, context-rich environments that support understanding even at low proficiency levels. It also contributes to reduced language anxiety by offering safe, engaging, avatar-mediated spaces for practice and experimentation. Creating activities rooted in sound SLA theories and principles becomes foundational when using VR, as "these technologies seem to be an 'ideal' instrument for language instruction, as they combine visual, auditory, and kinesthetic learning styles" (Scrivner et al., 2019, p. 55).

3. Evidence-based best practices

The following section describes evidence gathered from an experience we developed applying TBLT in a virtual environment. The activities highlight VR's potential for language acquisition through collaborative gaming, with a focused use of the target language (Italian), which requires completing relevant tasks. Students who study Italian are encouraged to consider studying abroad to reinforce their language skills and gain a deeper understanding of the culture through firsthand experience. However, not everyone can afford to do so. In addition to economic challenges, disabilities and other constraints can prevent students from traveling abroad (Bouchrika, 2024). By immersing students in their new virtual Italian apartment, VR offers a third space that moves them nearer to the study-abroad experience they would otherwise miss.

3.1. Classroom integration: a case study

The researchers conducted the study with 20 students at the upper-intermediate level of Italian. All students used Meta Quest 2 headsets, but the researchers also provided a non-tech alternative with paper-based activities. All students used Meta Quest 2 headsets, but the researchers also provided a non-tech alternative with paper-based activities to enhance accessibility options. Based on the notion of task and due to the immersive nature and kinesthetic qualities of virtual reality, the three virtual rooms (one for each activity) simulate tasks associated with studying abroad, such as furnishing one's bedroom by grabbing and moving objects and following commands.

The first virtual room reviews the target lexicon; the second experience asks a student to follow commands; and the final experience pairs two students to create the layout of their shared bedroom. While completing the tasks, students engage with a grammar structure (locational adverbs) and vocabulary related to bedroom furniture. The targeted adverbs include, for example, *under, on, and to the right,* among others, while the vocabulary words include *bed, nightstand, lamp,* and others.

The virtual experience begins in a furnished apartment, where an Italian roommate welcomes students and introduces them to their new home in Italy. In VRChat, user movement is either via teleportation or free movement. We recommend teleportation for beginners to reduce the risk of cybersickness.

Students start in a bedroom where an audio guide instructs them to drag Italian vocabulary labels onto the correct pieces of furniture. The room contains nine objects (Figures 9.1 and 9.2). Blocks labeled with the corresponding Italian words are available on nearby shelves. When a label is matched correctly to its object, students hear a positive response; if it is incorrect, they hear a negative response. Once all items are labeled, they follow arrows to a portal leading to the next room (Figure 9.3).

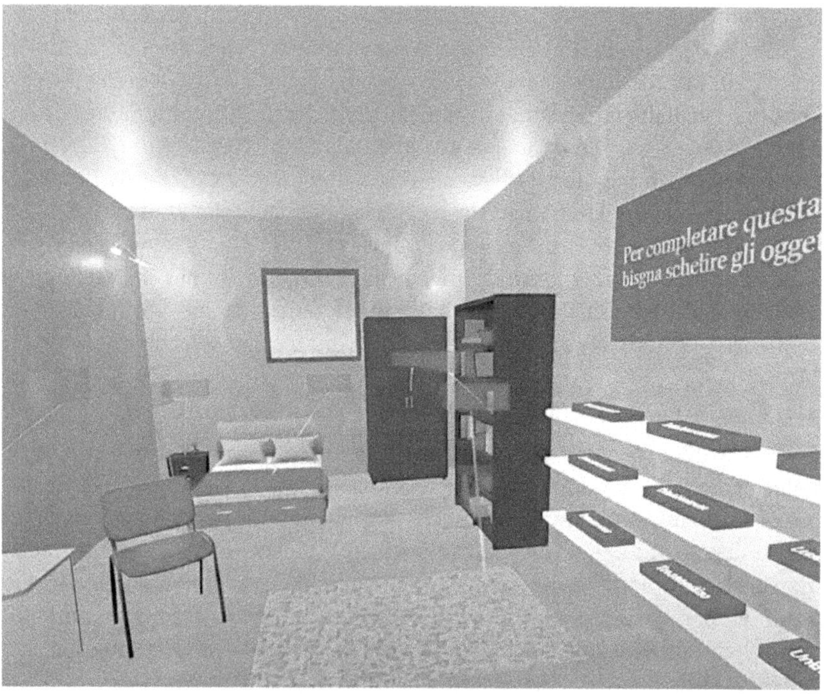

Figure 9.1. Furniture and labels in Room 1 (right side)

Figure 9.2. Furniture and labels in Room 1 (left side)

Figure 9.3. Portal to move to Room 2

In Room 2, students furnish a bedroom by following audio commands. The room is empty except for three reference points (a door, a window, and a poster) to support spatial orientation. Pressing a button on the wall triggers an audio instruction in Italian, prompting students to place the correct furniture item in a specified location. One object appears at a time in the center of the room, and the audio plays again if the item is dropped and picked up (Figure 9.4). Correct placements lock into place, and once students position all nine items, a voice confirms completion and triggers the appearance of another portal.

Figure 9.4. Grabbing and moving assets in Room 2

In Room 3, students collaborate in pairs to complete the same placement activity, this time using spoken Italian to direct one another. A brief instructional audio encourages them to take turns suggesting where to place the furniture using target-language phrases. Each pair works in a private version of the room generated upon entry. If students misplace objects because of miscommunication, partners are encouraged to correct each other in Italian, reinforcing comprehension and negotiation of meaning in a low-pressure, immersive environment (Figure 9.5).

Figure 9.5. Double set of assets for student interaction in Room 3

Defining the learner's level, as well as the correct quantity of input, cannot be calculated with precision. Gass (1997) noted that the notion of *comprehensible input* can be vague when the instructor attempts to define the level of knowledge a student possesses at a given time and to provide sufficient, appropriate input to acquire a new form. In class, the flow of activities from Room 1 to Room 3 builds upon preexisting knowledge and adds a layer of complexity. This layer of complexity is Krashen's *+1*, which increases from room to room. In Room 1, input consists of students hearing the name of the object when they select a label. In Room 2, input is the oral command to place assets in a designated location. In Room 3, input is the language produced by their partner.

Rooms 1 and 2 fulfill the goal of exposing students to comprehensible, meaningful input, as phrases are contextualized and relate to a larger task or situational purpose. The rooms also provide immediate, direct corrective feedback, allowing students to recognize a mismatch between their current state of knowledge and the target language. Even if students perform well on input activities (Rooms 1 and 2), it does not mean that the form has reached the level of *intake*. The goal is to guide students toward spontaneous production of utterances to communicate, as "there is a robust connection between interaction and learning" (Gass & Mackey, 2007, p. 176). For this reason, Room 3 is designed for controlled interaction, allowing students to use the structures and vocabulary to instruct each other on where to place a given asset. Therefore, Room 3 provides another form of input (this time given by the partner) and also the opportunity to produce controlled output. Output, even if controlled and limited as in Room 3, becomes essential, as it can confirm or challenge one's developing hypothesis about how language works and forces students to "process the morphosyntactic elements of language, as they attempt production, rather than just focus on semantic elements" (Loewen, 2020, p. 43).

The participants can request explicit feedback ("How do you say...?") and modify or reprocess the utterance based on the correction. Gass, Behney, and Plonsky (2013) consider feedback an indicator of students' success (or lack of success) and a crucial resource to focus on production. According to Swain (1993), it is necessary to give students as many in-class opportunities to write and speak as possible, primarily through peer-collaborative activities, which provide a "greater quantity of output and negotiation of meaning than that found in teacher-fronted classes" (p. 162). Peer corrective feedback is effective because it facilitates interactive scaffolding and creates a ZPD for learners. In the ZPD, learners receive support in mastering linguistic features they cannot yet manage independently (Sheen, 2010). This scaffolding enables learners to engage in the target language with the guidance of an interlocutor, leading to utterances that would be challenging to generate on their own. Although peer feedback is not guaranteed in interactive pair and group activities, an environment with a low affective filter—where students feel at ease—and strong motivation for immersion makes it more likely to occur.

3.2. Learner perspectives

To better understand how students experienced the intervention, all participants completed a post-activity questionnaire assessing enjoyment, technology use, modality comparisons, and perceived learning outcomes.

Although individual responses varied, often depending on students' prior exposure to digital tools and their comfort with VR, the data provide valuable insights into both the promise and the challenges of incorporating VR into language learning. Students rated their experience with the technology at an average of 6.4/10, with seven students giving high ratings. In addition, five students gave low ratings, explaining that the difficulty was also due to physical discomfort. Although only these five students had negative experiences due to cybersickness, 65% of all students reported uneasiness while in VR. Those who experienced mild to no cybersickness rated the experience 8/10, explaining that they appreciated learning in an immersive space, finding VR more effective than more traditional methods for acquiring and understanding the subject. These results suggest that while VR is enjoyable, prioritizing student wellness is crucial.

Most students (17 of 20) used the VR headsets during the in-class activity, while the others chose to complete the same tasks on laptops. Nine students had previously used a VR headset, though only occasionally (1–5 times), and half reported playing video games, typically only a few times per month. These results suggest that for many students, the intervention was among their first sustained experiences with immersive technology. This study showed that, on average, participants using VR rated the level of enjoyment of the activity at 7.6/10.

These findings highlight VR's promise as a tool for fostering engagement, embodied practice, and incidental language learning. Students consistently described VR as "fun," "cool," and "memorable," underscoring its potential to make language tasks more meaningful. At the same time, the data reveal the strong influence of usability challenges and physical discomfort (e.g., nausea, headaches, and navigation difficulties) which shaped the experiences of many participants. These results suggest that while VR can enhance task-based learning environments, careful attention must be given to technical support, session design, and student acclimation to ensure negative physical or technological experiences do not compromise learning outcomes.

3.3. Instructor preparation and scaffolding strategies

Before classroom implementation, the virtual rooms were beta-tested in weekly meetings by a group of ten instructors and collaborators, who explored the space, identified technical or pedagogical issues, and recommended improvements to enhance student engagement and learning. Most participants had little to no knowledge of the technology, allowing the team to predict barriers and address physical reactions. The researchers gradually adjusted headset time based on participants' responses to cybersickness. Such responses always led to the adoption of a seated position and a 5-minute time cap as the maximum duration before requiring a break.

Throughout the semester, the researchers implemented several scaffolding strategies to ensure students had a smooth experience with the technology. To further expose them to immersive environments, they watched a 360° video of an Italian town before the day of the intervention. This activity familiarized them with the use of controllers, wearing the headset, and experiencing virtual movement while seated. It also served to reduce potential anxiety and allowed the instructors to monitor students' reactions to VR.

Before the lesson focused on the virtual rooms, the researchers asked students to create a VRChat profile so they could log in promptly when using the headset. They also received a short instructional video via email, which guided them through the basic steps of navigating the platform and using the Meta Quest 2 headset. On the day of the intervention, a brief in-class presentation reviewed how to wear the headset, use the hand controllers, navigate the VRChat space, and interact with virtual objects and peers.

Since students in this course had previously encountered the targeted vocabulary and grammar points at the elementary level, they were only required to review the material at home and through online activities to refresh their knowledge. During the intervention, four instructors were present to provide the necessary linguistic and technological support for students as needed. The researchers instructed students to wear the headset for no more than five minutes and to pause earlier if they experienced any side effects. Students raised their hands if they needed assistance, and all four researchers were available to help.

These strategies not only facilitated a smooth transition into the virtual environment but also revealed essential challenges and solutions to consider when introducing students to new platforms or technologies. The following section will discuss the lessons learned and offer practical suggestions for educators.

4. Common challenges, limitations, and solutions

Despite the falling costs of hardware and the ever-increasing number of language and culture applications available, faculty members still face a range of challenges when incorporating VR into their curricula (Parmaxi, 2023). Common concerns include technical complexity, insufficient infrastructure, prohibitive costs, and issues related to curriculum integration, teacher preparation, or student accessibility (Ravichandran & Mahapatra, 2023; Wang & Li, 2024). In short, these concerns fall into three primary categories: (1) administrative (e.g., encompassing technical, institutional, and financial considerations); (2) instructor-focused (e.g., concerning teacher training and digital literacy); and (3) student-centered (e.g., addressing learner access and inclusivity). The following subsections identify common problems and provide targeted strategies and practical solutions for effectively integrating VR into language-learning curricula.

4.1. Technical, institutional, and financial considerations

Before introducing VR for language learning in the classroom, other technologies and methods should be considered and integrated into a comprehensive, multimodal implementation plan. VR provides the most immersive and context-rich environment for practicing spoken interactions and exploring cultural settings, but it is resource-intensive. Including non-tech or low-tech exercises can reduce this burden. At the same time, technologies such as AI-driven applications and digital storytelling can add variety and complement, introduce, or reinforce what students experience in the headset. Language applications like Duolingo, Babbel, and Memrise excel at gamifying a personalized practice with automated feedback, making them accessible though less contextualized. Likewise, digital storytelling tools such as *canva*, *genially*, and *voicethread* help students develop narrative skills, creativity, and reflective thinking through the use of multimedia expression. Together, these can form a comprehensive approach that combines the immersive power of VR with traditional methods, the accessibility of AI tools, and the reflective depth of storytelling, among other alternatives.

The second consideration is the specific technical prerequisites that must be budgeted for or funded through grants (Feridun & Bayraktar, 2024; Lee, Ray, Lai, & Tanner, 2022; Żammit, 2023). This funding can help with hardware, such as the purchase of high-quality VR headsets, the development or adaptation of pedagogically effective software, teacher training, and curriculum development. The most common forms of funding are federal and state grants, Ed-Tech programs, and corporate-sponsored initiatives.

Evaluating hardware is often the next step, and the Meta Quest mobile headset stands out due to its affordability, strong technical support, extensive app library, and ease of use. Other headsets, such as the Pico, Vive, or Index, attract attention as alternatives to the Meta Quest ecosystem but lag behind the Quest in at least one of the mentioned dimensions. Likewise, when identifying software, language learning-specific platforms like Immerse, Immerse Me, Engage, Mondly VR, Language Lab VR, and Victory XR (extended reality) are often chosen, as they offer ready-made scenarios, structured lessons, and built-in assessments for instructors to integrate into their lesson plans. Likewise, YouTube provides a substantial amount of 360^0 content ready to be utilized in a didactic manner. It is a common first step for educators who want to introduce VR in the classroom.

If an application, video, or service already aligns well with the planned pedagogical goals, instructors must also evaluate the technical infrastructure, such as Internet speed, reliability, and classroom space, to ensure it is suitable for use. Physical space requirements typically limit students to seated controller-based experiences, while at-home assignments could enable more natural movement. After selecting an experience, instructors load it onto the device. Fortunately, IT departments and staff at most schools assist with headset management and troubleshooting, reducing the burden on instructors. Updating and managing a fleet of headsets, however, can be time-consuming and typically requires a subscription to a management system, such as Meta for Education or Manage XR, for rapid deployment.

Educators who cannot find what they want within current platforms, as we did, can choose custom VR development. Given the typical instructor's workload, adopting streamlined workflows and efficient prototyping is essential. We recommend low-tech methods such as storyboards, dialogue maps, or simple mock-ups to clarify design concepts that focus on interactions before more intensive development begins. Tools like Unity and Unreal Engine lower technical barriers by offering accessible templates, visual scripting, and AI-assisted assets, animations, sprites, and code generation to help get bourgeoning developers started.

Unity is a beginner-friendly game and a common starting ground for VR developers because of its extensive documentation and support. It also offers a free VR development certification program through its learning platform Unity Learn. In addition to their own curriculum, there are numerous courses and tutorials available on YouTube, Coursera, Udemy, and other learning platforms, offering designers templates, beginner assets, and step-by-step instructions for building and understanding applications.

One of the features that Unity has over its competitors for VR development is the Meta All-In-One SDK (standard developer's kit), which includes its Building Blocks and Presence Platform for simple drag-and-drop prototyping of VR and MR applications. The Unity Asset Store also offers a diverse range of quick solutions at various price points. Alternative sources to the asset store for creators include open-source software like Blender and marketplaces such as Sketchfab (models), Mixamo (animations), and Freesound (audio), which, like other subscription-based (Synty Studios) and bundling (Humble Bundle) options, require a review of licensing and use agreements for publishing.

The simplest projects for beginners are those that limit perspectives to basic head movements, commonly referred to as 3-degrees of freedom (3-DOF), and these are often 360 ° film-based experiences. Platforms like ThingLink allow developers to include interactive, clickable elements, such as text boxes, videos, and audio playback, that users can view both in browsers and in headsets. More advanced projects integrate real or simulated body movement within virtual spaces and are known as 6-DOF experiences (Figure 9.6). Limiting the experience to a single user and leveraging templates, prefabricated assets, and AI tools like OpenAI's ChatGPT and Google's Gemini (for code and instructions) can simplify and accelerate the development process. The most complex projects to develop are multiplayer online environments; therefore, developers should build on established platforms such as VRChat, Rec Room, or Meta Horizon Worlds, which manage client-server connections.

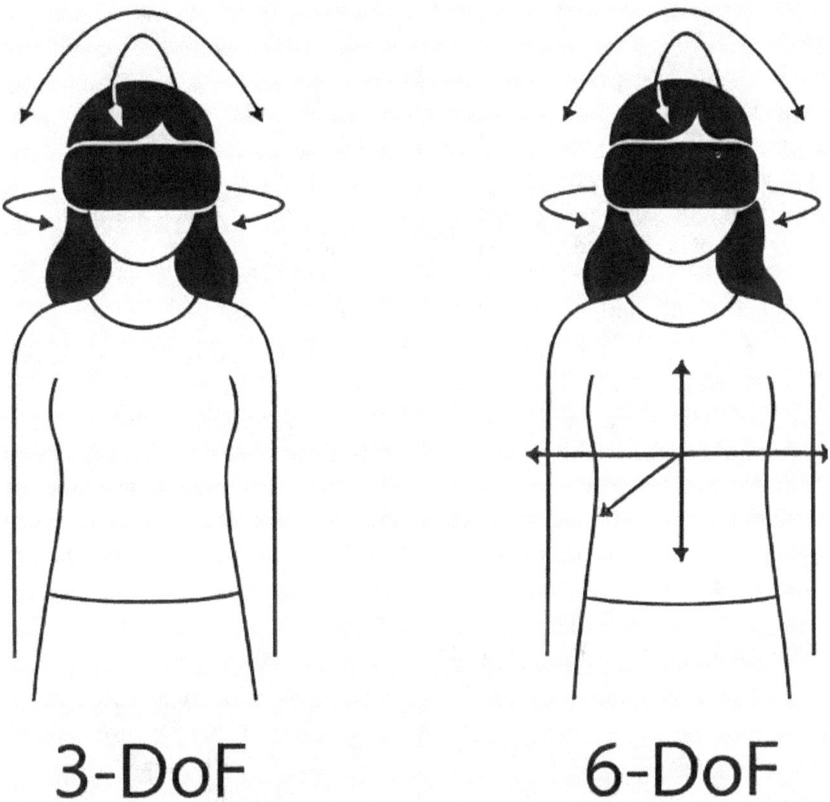

Figure 9.6. Six-degrees of freedom in VR

Note: OpenAI. (2025, July 3). Difference in motion tracking between 3DOF and 6DOF of freedom in VR headsets [AI-generated image]. ChatGPT. https://chat.openai.com/

For instructors who wish to develop more robust applications, additional costs may include access to a VR-ready computer, assets, or AI services, as well as personal time. While time-intensive, in-house development can significantly reduce the costs of software or services. When paired with other strategies, such as hardware cost-sharing, cloud-based infrastructure, and strategic corporate partnerships, implementing VR in the classroom could be both more accessible and widespread (Samala et al., 2025).

4.2. Teacher training and digital literacy

Aside from the technical concerns, teacher training and digital literacy have repeatedly been identified as critical factors in the adoption of VR for education (Lin & Wang, 2021; Liu et al., 2022). Instructors, while experts in pedagogy and their content area, may have limited experience with immersive technologies. For these teachers, working with VR can evoke both excitement and trepidation, as it balances the benefits of active learning against the time-consuming nature of the technology (Annamalai et al., 2023). This apprehension may also stem from digital literacy gaps that are more pronounced among tech-shy teachers who are skeptical or uncertain about the technology in general (Żammit, 2023). This uncertainty does not mean that they cannot become adept at using VR in their lessons. However, it does suggest that tailored professional development is needed to build confidence for this group through training, mentoring, and ongoing support.

Guidance would include orientation in the practical use of the headsets, VR-specific instructional design, examples of successful implementations, and the cultivation of a peer support network. Firsthand experience should involve demos where they can assess their headsets, learn how to launch and control applications, and manage common technical issues, such as tracking area calibration and restarting apps. Ready-made apps like Meta's First Steps, which demonstrates the headset setup and controls, or Wooorld, an app that integrates Google Earth's 3D and 360^0 content into a social space for encyclopedic and geographic group explorations, are but two applications that would both inform and inspire instructors in VR adaptability.

Further training should focus on helping interested instructors integrate VR into their planning, with thoughtful consideration given to how they will bookend the student experience with preparatory, scaffolding, alternative, and reflective activities that connect the virtual experience to their learning objectives. This approach encompasses knowledge of practical strategies for facilitating communication in a virtual environment, including screen-sharing and joining students in-world. This strategy also allows instructors to monitor student progress and behavior, and can lead to larger conversations on classroom norms, digital ethics, cyber safety, and respectful interactions.

Likewise, techniques such as the station rotation model, where small groups alternate between VR activities and complementary tasks, enable an instructor to manage the classroom with fewer headsets. Instructors should complement this work by fostering relationships that support peer learning and mentorship. In this regard, novices can pair with instructors who have already piloted VR in their classrooms, sharing insights, collaborating on lesson plans, and contributing to the growth of a community of practice.

4.3. Learner access and inclusivity

It is not enough for instructors to be trained and ready; for VR to become commonplace in the classroom, it must also be accessible and inclusive. Instructors must ensure that all students can engage meaningfully with the medium and that the assigned experiences support a diverse range of learners, including those who require accommodations or have special educational needs. If designers do not intentionally address this in the design process, VR could exacerbate educational inequities (Lee et al., 2020).

Some institutions have addressed this by setting up VR labs and equipment-lending libraries to make headsets available for at-home assignments. Extending access to students with special educational needs should begin earlier, as part of the hardware and software design process. For example, learners with visual impairments, mobility limitations, or sensory sensitivities may require alternative input methods, screen-reader compatibility, 3D audio, and customizable interfaces. Similarly, content should reflect a wide range of student backgrounds; otherwise, instructors risk alienating students who do not feel represented in the experience.

Inclusivity must also account for the physical well-being of students who may suffer from cybersickness, eye strain, and motion fatigue, particularly if sessions run too long. It is essential to limit initial VR use to short durations and provide alternatives (browser-based experiences, recordings, etc.) for students unable to participate directly. By applying universal design principles, training educators in VR pedagogy, and developing open-access resources, VR can become a transformative tool in language learning, offering rich and immersive experiences to students from all walks of life.

4.4. Areas for future consideration and research

The landscape of VR in language education is undergoing significant changes, primarily driven by rapid advances in AI tools and the miniaturization of hardware components. As a result, key areas for future consideration include experimenting with tools that can easily generate authentic environments, realistic AI-driven conversational agents, and the introduction of unobtrusive and mass-adopted mixed reality devices.

There already exist powerful tools like CityGen3D that can procedurally generate authentic environments. These applications can automatically generate detailed 3D city scenes from real-world map data. Automated scene generators like this one enable educators to quickly and inexpensively recreate real-world locations, from city streets to cultural landmarks, providing learners with immersive, context-rich settings that accurately reflect the target language's culture and environment.

Complementing these authentic environments is the development of AI-driven non-player characters (NPCs) that can engage learners in realistic conversations (in these environments). Subscription platforms like VictoryXR and Immerse are already offering AI-driven NPCs as tutors. At the same time, tools like Convai provide interfaces for developers to design their own language-specific NPCs with customizable voices, accents, vocabulary, phrases, and personality traits, thereby simulating more natural conversations with players. Notably, Meta continues to integrate their robust AI tools into their Horizon Worlds Desktop Editor for the creation of 3D assets, animations, sound effects, AI-driven NPCs, and text prompt-generated environments (Heaney, 2025). As these tools become more powerful and prevalent, they will need to be tested for their pedagogical value and to evaluate whether they provide accurate and culturally appropriate experiences, are free from bias, help students meet their learning objectives, and support social, emotional, and psychological well-being.

Another critical area for future planning concerns the evolution of hardware, particularly as VR headsets become smaller and smart glasses gain access to more computing power. The top investor in these technologies, Mark Zuckerberg, is betting that one of these two platforms will become the mass-adopted extended reality (XR) interface that replaces the smartphone (Bandara, 2024). This shift could expand language learning beyond classrooms into daily life by merging immersive simulations with real-world practice. The benefits and concerns are numerous, and these speculations would be better suited for another author.

While it is best to temper both the positive and negative anticipated effects, these technological advances are under development. They could provide the foundation for a future where educators can easily generate authentic virtual environments, populate them with virtual conversational partners, and have their students experience them on ubiquitous devices. Such innovations offer immense promises, but what is still needed is an evaluation of the impact of this technology on student outcomes that considers both their learning and social well-being. By doing so, the next generation of XR language learning may not only be excited with its novelty, but also deliver effective, evidence-backed outcomes.

5. Conclusion

Our experience with integrating VR into the intermediate-level Italian language classroom provided us with the opportunity to identify best practices and challenges, while also offering a compelling glimpse into the future of language education. However, there are many considerations when planning to create an application based on specific learning goals. It is necessary first to explore what already exists to meet these goals and incorporate this information when possible. When this is not the case, instructors may need to take development into their own hands, creating activities that are rooted in sound pedagogical theory and research. It is also important to remember that using VR for language learning is just one option, and it may not be effective for all instructors and students with different learning styles. Even so, VR language learning applications promote learner autonomy by encouraging self-direction, self-evaluation, and self-motivation. With constant improvements in more accessible VR platforms, assets, and other tools, users can achieve greater autonomy and increased efficiency in the future.

Nonetheless, we agree with McGrath et al. (2023), who state, whether and how AR and VR will fundamentally influence research, teaching, and learning in the next decade is still up in the air. Seeking a crystal-ball view of augmented reality (AR) and VR product adoption for academic technology planning purposes would be unavailing." Despite the hurdles, the experience shared in this chapter should contribute valuable insights to the growing body of research on immersive learning and for those teaching with this technology. It also highlights the need for more accessible tools, shared resources, and professional development opportunities to empower educators to experiment with and adopt VR in meaningful ways.

References

Ahmadian, M. J., & Long, M. H. (2021). *The Cambridge handbook of task-based language teaching.* Cambridge University Press. https://doi.org/10.1017/9781108868327

Allcoat, D., & von Mühlenen, A. (2018). Learning in virtual reality: Effects on performance, emotion, and engagement. *Research in Learning Technology, 26,* 1–13. https://doi.org/10.25304/rlt.v26.2140

Annamalai, N., Uthayakumaran, A. & Zyoud, S.H. (2023). High school teachers' perception of AR and VR in English language teaching and learning activities: A developing country perspective. *Education and Information Technologies, 28,* 3117–3143. https://doi.org/10.1007/s10639-022-11275-2

Bandara, P. (2024, December 31). *PetaPixel.* Retrieved March 21, 2025, from https://petapixel.com/2024/12/31/meta-is-reportedly-planning-to-add-display-screens-to-its-smart-glasses

Bouchrika, I. (2024, February 8). *Studying abroad in 2024: Reasons, challenges & tips.* Retrieved April 20, 2025, from https://research.com/education/studying-abroad

Dede, C. (2009). Immersive interfaces for engagement and learning. *Science, 323*(5910), 66–69. https://doi.org/10.1126/science.1167311

Dooly, M., Thrasher, T., & Sadler, R. (2023). "Whoa! Incredible!": Language learning experiences in virtual reality. *RELC Journal, 54*(2), 321–339. https://doi.org/10.1177/00336882231167610

Ellis, R., Skehan, P., Li, S., Shintani, N., & Lambert, C. (2019). *Task-based language teaching: Theory and practice.* Cambridge University Press. https://doi.org10.1017/9781108643689

Feridun, K. B., & Bayraktar, Ü. (2024). The future of virtual reality and education. *Turkish Online Journal of Educational Technology, 23*(3), 110–119.

Gass S. M. (1997). *Input, interaction, and second language learner.* LEA. https://doi.org/10.4324/9780203053560

Gass, S. M., Behney, J., & Plonsky, L. (2013). *Second language acquisition: An introductory course.* Routledge. https://doi.org/10.4324/9780203137093

Gass, S. M., & Mackey, A. (2007). Input, interaction, and output in second language acquisition. In B. VanPatten & J. Williams (Eds.), *Theories in second language acquisition* (pp. 175–200). LEA.

González-Lloret, M., & Ortega, L. (2014). Towards technology-mediated TBLT: An introduction. In M. González-Lloret & L. Ortega (Eds.), *Technology-mediated TBLT: Researching technology and tasks* (pp. 1–22). John Benjamins. https://doi.org/10.1075/tblt.6

Graeske, C., & Sjöberg, A. S. (2021). VR technology in teaching: Opportunities and challenges. *International Education Studies, 14*(8), 76–83. https://doi.org/10.5539/ies.v14n8p76

Heaney, D. (2025, June 18). *Meta teases AI generation of entire Horizon Worlds coming "very soon."* UploadVR. Retrieved May 12, 2025 from https://www.uploadvr.com/meta-teases-ai-generation-of-entire-horizon-worlds-coming-very-soon

Horwitz, E. K. (1988). The beliefs about language learning of beginning university foreign language students. *The Modern Language Journal, 72*(3), 283–294. https://doi.org/10.2307/327506

Horwitz, E. K. (2001). Language anxiety and achievement. *Annual Review of Applied Linguistics, 21,* 112–126. https://doi.org/10.1017/S0267190501000071

Horwitz, E. K., Horwitz, M. B., & Cope, J. (1986). Foreign language classroom anxiety. *The Modern Language Journal, 70*(2), 125–132. https://doi.org/10.2307/327317

Jackson, D. O. (2022). *Task-based language teaching.* Cambridge University Press. https://doi.org/10.1017/9781009067973

Johnson-Glenberg, M. C., Bartolomea, H., & Kalina, E. (2021). Platform is not destiny: Embodied learning effects comparing 2D desktop to 3D virtual reality STEM experiences. *Journal of Computer Assisted Learning, 37*(5), 1263–1284. https://doi.org/10.1111/jcal.12567

Krashen, S. (1985). *The input hypothesis: Issues and implications.* Longman.

Lee, N., Ray, R., Lai, S., & Tanner, B. (2022). *Ensuring equitable access to AR/VR in higher education.* Brookings Institution. Retrieved May 12, 2025, from https://www.brookings.edu/articles/ensuring-equitable-access-to-ar-vr-in-higher-education

Lin, Y. J., & Wang, H. C. (2021). Using virtual reality to facilitate learners' creative self-efficacy and intrinsic motivation in an EFL classroom. *Education and Information Technologies, 26,* 4487–4505. https://doi.org/10.1007/s10639-021-10472-9

Liu, R., Wang, L., Koszalka, T. A., & Wan, K. (2022). Effects of immersive virtual reality classrooms on students' academic achievement, motivation, and cognitive load in science lessons. *Journal of Computer Assisted Learning, 38*(5), 1422–1433. https://doi.org/10.1111/jcal.12688

Loewen, S. (2020). *Introduction to instructed second language acquisition* (2nd ed.). Routledge. https://doi.org/10.4324/9781315616797

Long, M. H. (1996). The role of the linguistic environment in second language acquisition. In W. Ritchie & T. Bhatia (Eds.), *Handbook of second language acquisition* (pp. 413–468). Academic Press.

Long, M. H. (2015). *Second language acquisition and task-based language teaching.* Wiley-Blackwell. ProQuest Ebook Central. https://ebookcentral-proquest-com.proxy.lib.ohio-state.edu/lib/ohiostate-ebooks/detail.action?docID=7104179

McGrath, O., Hoffman, C., & Dark, S. (2023, April 20). Future prospects and considerations for AR and VR in higher education academic technology. *Educause Review.* Retrieved May 25, 2025, from https://er.educause.edu/articles/2023/4/future-prospects-and-considerations-for-ar-and-vr-in-higher-education-academic-technology

Nunan, D. (2015). *Task-based language teaching.* Cambridge University Press.

Parmaxi, A. (2023). Virtual reality in language learning: A systematic review and implications for research and practice. *Interactive Learning Environments, 31*(1), 172–184. https://doi.org/10.1080/10494820.2020.1765392

Pataquiva, F. de P. F., & Klímová, B. (2022). A systematic review of virtual reality in the acquisition of a second language. *International Journal of Emerging Technologies in Learning, 17*(15), 43–53. https://doi.org/10.3991/ijet.v17i15.31781

Ravichandran, R. R., & Mahapatra, J. (2023). Virtual reality in vocational education and training: Challenges and possibilities. *Journal of Digital Learning and Education, 3*(1), 25–31. https://doi.org/10.52562/jdle.v3i1.602

Samala, A. D., Rawas, S., Rahmadika, S., Criollo-C, S., Fikri, R., & Sandra, R.P. (2025). Virtual reality in education: Global trends, challenges, and impacts—Game changer or passing trend? *Discover Education, 4,* 229. https://doi.org/10.1007/s44217-025-00650-z

Schmidt, R. (1990). The role of consciousness in second language learning. *Applied Linguistics, 11*(2), 129–158. https://doi.org/10.1093/applin/11.2.129

Smith, B., & González-Lloret, M. (2021). Technology-mediated task-based language teaching: A research agenda. *Language Teaching, 54*(4), 518–534. https://doi.org/10.1017/S0261444820000233

Smith, M. D., & McCurrach, D. (2021). The usage of virtual reality in task-based language teaching [Conference presentation]. *KOTESOL International Conference Proceedings, 2021.* https://doi.org/10.5281/zenodo.5595431

Swain, M. (1985). Communicative competence: Some roles of comprehensible input and comprehensible output in its development. In S. M. Gass & C. Madden (Eds.), *Input in second language acquisition* (pp. 91–103). Newbury House.

Swain, M. (1993). The output hypothesis: Just speaking and writing aren't enough. *The Canadian Modern Language Review, 50*(1), 158–164. https://doi.org/10.3138/cmlr.50.1.158

Tai, T., Howard, H. C., & Todd, G. (2020). The impact of a virtual reality app on adolescent EFL learners' vocabulary learning. *Computer Assisted Language Learning, 32*(4), 892–917. https://doi.org/10.1080/09588221.2020.1752735

Uebiyev, D., & Facer, B. R. (2025, June 25–28). *Virtual worlds, real insights: Exploring student reflections from ImmerseMe* [Conference presentation]. FLEAT 8 Conference. https://doi.org/10.13140/RG.2.2.11236.08320

Vygotsky, L. S. (1978). *Mind in society: The development of higher psychological processes.* Harvard University Press. https://doi.org/10.2307/j.ctvjf9vz4

Wang, Q., & Li, Y. (2024). How virtual reality, augmented reality, and mixed reality facilitate teacher education: A systematic review. *Journal of Computer Assisted Learning, 40*(3), 1276–1294. https://doi.org/10.1111/jcal.12949

Żammit, J. (2023). Exploring the effectiveness of virtual reality in teaching Maltese as a second language. *Computers & Education: X Reality, 3,* 100035. https://doi.org/10.1016/j.cexr.2023.100035

Chapter 10

Enhancing second language learning with captions

Wing Shan Chan
National Coalition of Independent Scholars

S. M. Sabra
University of California, Irvine

Abstract

Multimodal learning has become an essential learning environment for online education. Captioned video as an audiovisual input is a flexible learning tool for second language learning. This chapter explores the strategic use of caption types in second language learning. Integrating four second language acquisition theories, the mechanism of second language learning is explored based on the theoretical foundations of dual coding and cognitive load theory. Existing studies have shown that various caption types such as L2 captions, L1 subtitles, and bilingual subtitles can enhance second language learning in distinct ways depending on learners' proficiency levels. Findings have consistently evidenced that L2 captions facilitate more challenging authentic language comprehension in advanced learners while L1 subtitles aid in the understanding of learners with less proficiency. Bilingual subtitles may offer greater benefits in meaning recognition during vocabulary learning and content retention. Different caption enhancement forms such as keyword, highlighted, and glossed captions, can facilitate language comprehension and vocabulary development by directing learners' attention to salient linguistic features and providing contextual support. A dynamic approach is proposed that integrates interaction between learner characteristics, learning contexts, and instructional objectives when implementing captioned videos for second language learning. This chapter offers practical guidelines and considerations to language instructors in utilizing captions as part of their teaching instructions for language learning and those committed to implementing effective pedagogical practices. Recommendations are also offered for further research on personalized captioning systems and advanced technology integration which could optimize language learning outcomes.

Keywords: Second language acquisition, Subtitles, Audio-visual materials; human information processing; Educational technology

* * *

1. Introduction: captions in second language acquisition

Since the COVID-19 pandemic, digital multimodal learning has become a prominent learning method. A multimodal learning environment presents information in different modes (verbal and nonverbal), engaging multiple sensory systems such as auditory and visual channels. Multimodal learning has become a significant educational environment for students globally, including second language learners (Bao, 2020).

Video with captions as an audiovisual input has become an essential tool in second language acquisition, offering authentic input and multimodal learning opportunities enabled by technological advancements and accessible digital media. The information processing frameworks support the integration of captions in second language learning and align with second language acquisition theories. Captioned videos synchronize written text with spoken dialogues, helping learners to establish links between spoken and written forms to support deeper comprehension of authentic audiovisual materials (Danan, 2004; Gass et al., 2019; Montero Perez, 2022). Different caption types and enhancement forms provide flexible pedagogical strategies that can be tailored to learners' proficiency level and learning objectives, making captions an invaluable tool for second language learning.

The terms captions and subtitles are sometimes used interchangeably but often refer to different types of on-screen text. Captions refer to the on-screen text that is presented in the same language as the spoken dialogues—in this case, learners' second language (L2). On the other hand, subtitles refer to the translation of the L2 spoken audio into learners' first language (L1). Bilingual subtitles refer to the presentation of both L2 captions and L1 subtitles at the same time in two separate lines (Montero Perez, 2022).

Partially adopted from a doctoral dissertation, this chapter explores how strategic use of caption types can support comprehension, vocabulary development, and learner autonomy across proficiency levels.

2. Theoretical foundation in second language acquisition

Information processing frameworks view the mind as a system that encodes and stores information, using mental structures to process sensory inputs into cognitive outputs (Sweller, 2011, 2024). This section discusses dual coding theory (Paivio, 1991) and cognitive load theory (Sweller, 2011, 2024) to deepen understanding of how second language learners process on-screen text presented as L2 captions and L1 subtitles. Both theories acknowledged the limitations of working memory and offered strategies to maximize learning within the constraints of human cognitive resources. Through the lens of second language acquisition theories, the implementation of captioned videos reflects the principles of comprehension input theory (Krashen, 1982), noticing hypothesis (Schmidt, 1990), output hypothesis (Swain, 1985), and task-based learning (Ellis, 2000), underscoring the pedagogical value of caption use as a strategic and theory-informed tool for enhancing second language learning.

2.1. Dual coding theory

Dual coding theory (Paivio, 1991) proposes that learners process verbal information (e.g., speech, music) and nonverbal information (e.g., written text, images) simultaneously through auditory and visual channels in the human cognitive system. Dual coding theory assumes that both verbal and non-verbal systems are functionally independent yet interconnected. The interconnectedness of the systems enables information to be accessed and activated during the process of learning and comprehension through verbal or visual cues (Figure 10.1).

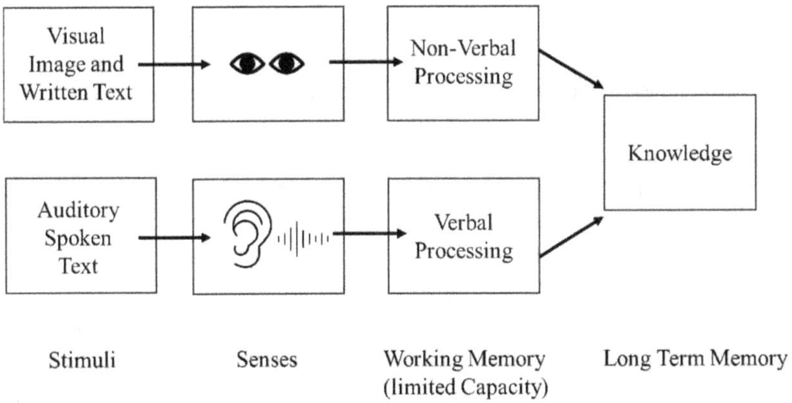

Figure 10.1. Illustration of information processing in dual coding theory

For example, in vocabulary learning, as shown in Figure 10.2, a written word (e.g. cat) can be presented simultaneously with its equivalent visual image (e.g. a picture of a cat) and its correct pronunciation (spoken word of "cat"). Learners will then form an association between these representations in different modalities (visual text, visual image, auditory text) associated with the concept "cat". Based on dual coding theory, showing learners the picture of a cat would activate the visual memory of the word "cat" and its verbal pronunciation through previous association. Consequently, the word "cat" is learned if learners (1) have successfully formed the correct association of those mental representations, (2) can recognize the word "cat" from the visual system, and (3) are able to say the word "cat" correctly when seeing the cat picture.

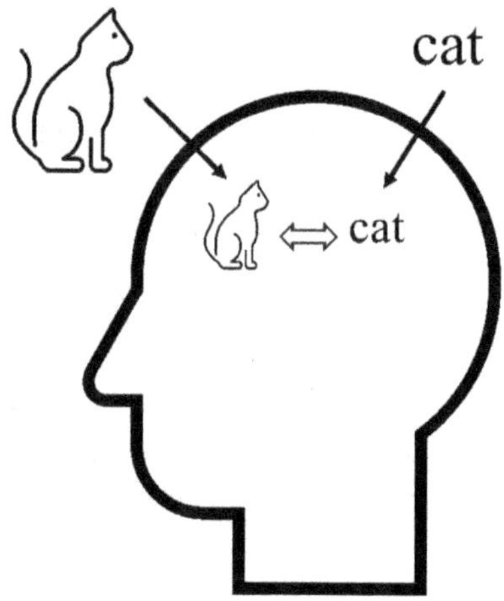

Figure 10.2. Vocabulary learning in dual coding theory

As learners become more well-learned, the interconnected network between the systems becomes more extensive and intricate. The process also becomes more complex when presented with unfamiliar and complicated information. In the case of second language learning with captioned videos, seeing and hearing the same message reinforces retention and strengthens mental connections (Paivio, 1991).

2.2. Cognitive load theory

Cognitive load theory has been used frequently to understand the cognitive processing of on-screen text presented as captions and subtitles. Developed by Sweller (2011, 2024) with the inception to generate instructional procedures for effective learning, cognitive load theory is a theoretical construct that is widely used in instructional designs and educational research. The theory is based on two major assumptions in which human cognitive architecture consists of an unlimited long-term memory and a limited working memory. Long-term memory coordinates human cognitive activities whereas working memory processes new information and transfers the new learned information to be stored in long-term memory.

Effective cognitive processing depends largely on the working memory capacity and its ability to automatically retrieve prior knowledge from long-term memory when processing novel information (Baddeley, 1992). Consequently, complex tasks with ambiguous instructions or unfamiliar information impose high extraneous load, which overwhelms learners' working memory and thus impedes learning.

In video-based second language learning, the presence of captions reduced extraneous cognitive load, allowing learners to use less working memory to process unfamiliar sounds, which enables them to interpret meaning and grammar with ease (Sweller, 2011). This also explains why language proficiency has been shown as a major determining factor for effective captions (Danan, 2004; Garza, 1991; Markham, 1999; To, 2024; Vanderplank, 1988). High language proficiency indicates a good command of linguistic skills that have been learned and stored in long-term memory. These language skills are ready to be retrieved for effective processing in working memory when reading on-screen text, allowing more cognitive capacity to process new information and leading to better comprehension.

2.3. Second language acquisition theories

The implementation of captioned videos in second language learning can be effectively contextualized through established second language acquisition theories, as illustrated by the progressive stages of language learning in the following. Comprehensible input theory (Krashen, 1982) explains that language learning occurs when learners are exposed to language inputs that are slightly above their current proficiency. Therefore, matching proficiency level with captions makes input understandable and accessible (Gass et al., 2019; Pujadas & Muñoz, 2024). Second language acquisition theories stated that the mere exposure to language input is not enough; deliberate noticing of linguistic features and elements is essential in enhancing learners' awareness that might not be consciously perceived (Lee & Révész, 2020). The strategic application of caption types and enhancement forms is informed by the principles of the noticing hypothesis (Schmidt, 1990), which claims that successful learning of specific linguistic features such as grammar and vocabulary can only be achieved by explicit noticing.

The ability to produce a language is equally important for language development as the ability to understand it. Output hypothesis theory (Swain, 1985) emphasizes the importance of learners producing language on their own, allowing them to notice gaps in their knowledge for self-correction and reflective learning. Output hypothesis promotes language use and acquisition through task-based activities, creating authentic real-world experience of language use. In task-based learning (Ellis, 2000), captions can be used as a resource via learner-directed tasks, such as summarizing, grammar exercises, or reenacting video content, prompts learners to actively engage with the language, promoting language learning and comprehension (Lee & Révész, 2020).

3. Research evidence on captioning

Over the past two decades, a wealth of empirical research has demonstrated the positive impact of instructional strategies on language learning using authentic videos with L2 captions. Studies spanning from Garza (1991) and Vanderplank (1988) to more recent works by Aldukhayel (2021), Teng (2020), Lo (2024), and Pattemore et al. (2024) have demonstrated similar positive findings. These findings consistently show that while L2 captions facilitate comprehension in advanced learners, L1 subtitles support the understanding of less proficient learners. While the review of Wei and Fan (2022) found that L2 captions generally outperform L1 subtitles in vocabulary learning, the mixed result across studies suggests that the effectiveness of caption types may be influenced by variables such as learner proficiency, choice of caption types, and contextual factors.

3.1 Learning outcomes benefited from captions

Vocabulary is the initial building blocks of learning a new language. The implementation of captions promotes vocabulary development by providing repeated and meaningful exposure to words and phrases in context. Research shows that captioned videos lead to higher vocabulary gains compared to uncaptioned ones, and that L2 captions are particularly effective in facilitating vocabulary development (Kurokawa et al., 2024; Teng, 2020;).

Listening comprehension is a primary challenge for second language learners. Rapid connected speech and unfamiliar accents are particularly demanding. Captions offer visual support by segmenting continuous speech, clarifying word boundaries, and linking phonological and orthographic forms (Garza, 1991; Montero Perez et al., 2013). The display of captions alongside spoken language reinforces recognition, spelling, form-meaning mapping, comprehension, and retention (Montero Perez et al., 2018; Pattemore & Muñoz, 2020; Pujadas & Muñoz, 2024).

Captions also facilitate learner autonomy by giving them the ability to control their level of engagement with the video content, allowing them to pause, replay, or adjust caption settings. Through self-monitoring, learners can gauge their language proficiency and gradually transition between different caption types (e.g., L1 to L2 captions), eventually reducing the use of captions entirely. This progression fosters autonomy, confidence, and the development of real-world listening skills (Pujadas & Muñoz, 2024). This control builds learners' confidence and motivation while supporting more independent language use.

3.2. L2 captions: impact on advanced learners

Captions draw learners' attention to new lexical phenomena, which increases the likelihood and ability of vocabulary learning (Lee & Révész, 2020). Most importantly, advanced learners' proficiency allows them to process and integrate both visual and auditory modalities effectively without overloading their cognitive processing (Aldukhayel, 2021; Garza, 1991; Muñoz et al., 2022; Pattemore & Muñoz, 2020).

L2 captions improve listening comprehension for advanced learners, particularly when the content is challenging but not overwhelming. The benefit is most salient when captions are matched to the learner's proficiency and cognitive load is manageable (Montero Perez et al. 2013; Pujadas & Muñoz, 2024). L2 captions were also found to support speech segmentation development, especially for learners with high proficiency. Speech segmentation is a crucial part of the listening process that identifies the boundaries between individual words in a continuous stream of spoken speech (Hsieh, 2019; Montero Perez, 2022). Moreover, captions assist advanced learners in processing ambiguous input, such as unfamiliar accents or rapid speech (Markham, 1999).

However, the impact of L2 captions diminishes as proficiency increases. Previous studies suggest that the expertise reversal effect may occur when captions divert learners' attention from natural language processing, thereby reducing learning effectiveness (Muñoz et al., 2023; Pattemore & Muñoz, 2020; Sweller, 2024).

3.3. L1 subtitles: support for beginners

Beginner learners benefit most from L1 subtitles, which support deeper cognitive processing and lead to improved comprehension and retention of content (Chan et al., 2022; Danan, 2004; Markham, 2001; To, 2024; Wi & Boers, 2023). Assuming learners with native language reading competence, L1 subtitles can serve as a bridge between languages to access meaning, making L2 audiovisual input accessible. L1 subtitles reduce cognitive overload, lower anxiety, and build confidence, allowing learners to gradually transition to L2 captions (Montero Perez, 2022; To, 2024).

3.4. Bilingual subtitles: vocabulary and retention

Empirical evidence has shown that viewing bilingual subtitled videos can foster vocabulary learning as learners directly associate new words and phrases with their meanings in real-time, leading to better comprehension and retention (Lo, 2024). Bilingual subtitles significantly improve students' L2 vocabulary recognition and recall compared to monolingual subtitles in videos. Wang and

Pellicer-Sánchez (2022) highlighted that bilingual subtitles are a valuable tool that can create incidental vocabulary learning. Incidental vocabulary learning refers to unintentional learning of new words through watching captioned videos while focusing on understanding video content. Their eye-tracking results indicated that bilingual subtitles offered greater benefits as opposed to L2 captions or L1 subtitles in meaning recognition, whereas L2 captions is more effective in establishing form recognition.

The eye-tracking results by Liao et al. (2020) showed that beginner learners paid more attention to L1 subtitles when bilingual subtitles were presented, the findings still indicated the positive effect of bilingual subtitles. This suggests that bilingual subtitles could be an effective tool in language education, providing a manageable approach to understanding and retaining new vocabulary according to learners' needs. However, there is insufficient data on the effect of bilingual subtitles, making it difficult to draw firm conclusions about their role in vocabulary acquisition.

3.5. Gaps in the research

Despite growing research in captioned videos as a tool for language learning, several research gaps remain. While studies show that L2 captions are more effective in promoting vocabulary learning, the evidence surrounding bilingual subtitles is limited and inconclusive. This lack of data makes it difficult to assess their comparative effectiveness or understand how learners interact with bilingual input. Most studies focus on English as the target language with learners speaking Indo-European language, more research is needed for other language pairs (Wei & Fan, 2022). In terms of research design, the use of eye-tracking methodology can provide objective perspective of how on-screen texts are read and processed. Other learner variables such as proficiency level, cognitive load, and individual differences, especially among neurodivergent learners, are also underrepresented in current research. Additionally, methodological inconsistencies across studies, such as variations in caption design, vocabulary measurement, and exposure time, further complicate efforts to synthesize findings. Addressing these gaps could lead to more insights and better-informed pedagogical strategies for integrating captioned videos into language learning environments.

4. Practical implementation strategies

Determining learners' language proficiency is crucial to selecting appropriate types of captions (e.g., L1 subtitles, L2 captions, bilingual subtitles) for second language learning using captioned videos. Matching caption types and their enhancement forms to learners' proficiency and learning goals can greatly improve learning outcomes.

4.1. Caption types and learner proficiency

Existing studies demonstrated that advanced learners benefited from L2 captions while beginners benefited most from L1 subtitles (Chan et al., 2022; Danan, 2004; Markham, 2001; To, 2024; Wei & Fan, 2022) in language learning. According to cognitive load theory mentioned earlier, high language proficiency enables efficient retrieval of stored skills, freeing cognitive resources for processing new information and enhancing comprehension and overall language skills when reading L2 captions. Since both spoken information and captions are transient and disappear immediately, beginner learners lack the L2 competence and reading speed to process audiovisual input through listening and reading, which requires memorizing information for processing (Leahy & Sweller, 2011). On the other hand, advanced learners comprehend target content more effectively by utilizing various information sources, including captions and visual cues (Chan et al., 2022).

The eye-tracking study by Chan et al. (2022) provides visual evidence (unpublished) of how Chinese-speaking learners distributed their on-screen attention in the presence of more proficient language (Chinese L1 subtitles) and less proficient ones (English L2 captions). Even though this study focused on learning through learners' second language, the findings offer insight into the relationship between caption types and learners' proficiency. Each line shown in Figures 10.3 and 10.4 represent a visual scan path for one participant. A large circle means more time was spent looking at the area.

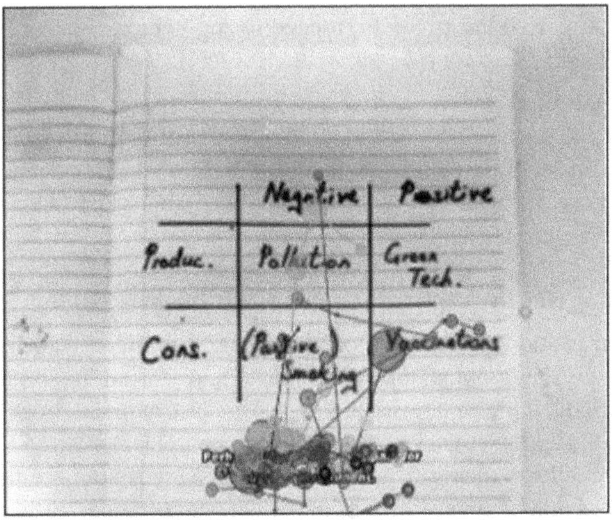

Figure 10.3. Scan paths of learners reading on-screen text in a less proficient language (L2 captions)

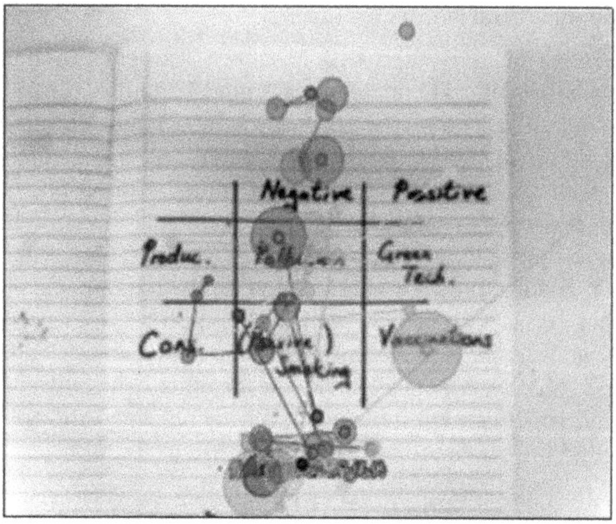

Figure 10.4. Scan paths of learners reading on-screen text in a more proficient language (L1 subtitles)

The findings of Chan et al. (2022) reveal that Chinese-speaking learners spent more time reading L2 captions, which are on-screen text in a less proficient language, due to the unfamiliar vocabulary (Figure 10.3). Conversely, these learners allocated less time reading L1 subtitles and focused more on other areas of the screen (Figure 10.4). Their subtitle reading patterns indicate that L1 subtitles supported information processing and comprehension by reducing cognitive load.

The implication of this eye-tracking study illustrates that using learners' proficient language as on-screen text, that is L1 subtitles in second language learning, can reduce cognitive load, and therefore facilitate deeper cognitive processing in comprehending L2 information. This is particularly true for beginner learners in accessing L2 content.

4.2. Creating effective captions

To optimize captioned videos in second language learning, selecting appropriate caption types is essential for successful language learning through digital platforms. Full captions are usually expected in captioned videos; however, variations of caption enhancement forms can be implemented to enhance the pedagogical effectiveness of captioned videos, supporting more targeted and impactful second language learning, particularly for beginner learners.

Understanding how to create captions helps educators develop effective audiovisual resources for language learning. While platforms such as YouTube offer auto-captions generated by speech recognition and artificial intelligence (AI), free video editing applications such as CapCut enables animated text and color options which are easy to create (Roshne et al., 2024).

Creating captions that are visually salient within the caption line can attract learners' attention. Features such as keyword captions, glossed captions, highlighted captions, and full captions with highlighted target words are examples of strategies used to achieve this (Aldukhayel, 2021; Teng, 2020). These enhancements are designed to help learners focus on important or difficult words while watching videos, making it easier to notice and learn new vocabulary. Visually salient captions and enhancements are in line with noticing hypothesis (Swain, 1985) to improve learning outcomes by drawing learners' attention and encouraging explicit noticing (Teng, 2020).

Figure 10.5 shows an example of glossed caption function that is available through a browser extension downloaded from Language Reactor (https://www.languagereactor.com/), an online language learning platform that enables the videos shown on Netflix or YouTube with glossary feature. Viewers can move their cursor over any particular words of the subtitle to see its equivalent translation and/or meaning. This feature includes vocabulary pronunciation and links to external sites (Figure 10.6), if viewers need further information about the words.

Figure 10.5. Example of glossed captions using a browser extension downloaded from Language Reactor website.

Note: The screenshot was taken from a video that forms part of an ongoing language project in which the author is involved. Permission for its use was granted by the Cantonese Alliance.

Figure 10.6. Links to external sites (circled area) are available if learners need more information regarding a particular vocabulary.

Note: The screenshot was taken from a video that forms part of an ongoing language project in which the author is involved. Permission for its use was granted by the Cantonese Alliance.

Bajorek (2017) discussed how this feature provides pronunciation guides, which helped learners mimic native speech patterns, allowing them to improve speaking and listening skills alongside text. The caption feature can significantly boost language learning outcomes by providing immediate, context-rich feedback that helps learners correct their mistakes and understand language nuances on the spot. However, the availability of such information depends on what is available across the Internet.

Table 10.1 identifies the different caption types and their benefits for learners at varying proficiency levels, including suggested classroom implementation.

Learning Outcome	Proficiency	Caption Types	Suggested Classroom Caption-Based Activities
Vocabulary	Beginner	1. Keyword L2 Captions 2. Highlighted L2 Captions 3. Full L2 Caption with Highlighted Target-Word 4. Glossed Captions 5. Bilingual Subtitles	1. Create fill-in-the-blank questions or comprehension questions. 2. Pre-learning new vocabulary. 3. Watched video with visually enhanced captions for targeted vocabulary. 4. Teacher explains and class discussion of the video. 5. Watched the captioned video the second time, paused the video to ask the fill-in-the-blank questions or comprehension questions. Can be answered orally or in writing. *This task-flow can be adopted flexibly and creatively for other caption type activity.
Vocabulary	Advanced	1. L2 Captions 2. Full L2 Caption with 3. Highlighted Target-Word 4. Bilingual Subtitles	1. Use or create visually enhanced captions for new or difficult vocabularies. 2. Recall video content orally or in writing. 3. Create new sentences using newly learnt vocabularies. 4. Create stories using newly learnt vocabularies.
Listening Comprehension	Advanced	L2 Captions	1. Sequential learning: L2 captions => no caption 2. Recall video content orally or in writing. 3. Transcribe spoken dialogues

Comprehension	Beginner	1. L1 Subtitles 2. Bilingual Subtitles	1. Sequential learning: L1 subtitles => L2 captions / bilingual subtitles => no caption 2. Implement glossed captions.
Comprehension	Advanced	1. L2 Captions 2. Highlighted Captions 3. Bilingual Subtitles	1. Sequential learning: L2 captions => no caption 2. Recall video content orally or in writing. 3. Highlighting captions for new linguistic expressions or features. 4. Create captions for video clip
Segmentation	Beginner	Highlighted L2 Captions	1. Highlighting captions for new linguistic expressions or features. 2. Fill-in-the-blank questions and comprehension questions. 3. Speak aloud.
Segmentation	Advanced	L2 Captions	1. Comprehension questions 2. Recall video content 3. Multiple choice questions
Grammar	Beginner	Keyword / highlighted L2 Captions	1. Composition using targeted linguistic elements. 2. Fill-in-the-blank questions 3. Multiple choice questions
Grammar	Advanced	L2 Captions	1. Composition using targeted linguistic elements. 2. Fill-in-the-blank questions 3. Multiple choice questions

Table 10.1. Various caption types and their benefits for learners at various levels of proficiency.

4.3. Designing caption-based activities

Caption-based activities provide linguistic support during listening or watching captioned video, allowing learners to engage more actively and confidently. Teachers can incorporate keywords or captions highlighted videos to train learners to read efficiently and focus on key information, thereby improving comprehension despite the fast pace of authentic speech. As learners progress in their L2 development, their reliance on L2 captions will gradually reduce (Hsieh, 2019).

Introducing new vocabulary or linguistic elements during pre-learning activities before watching captioned videos improves vocabulary learning. Montero Perez (2019) explained that this approach stimulates form-meaning mapping of targeted words before video watching, directing attentional resources to learning targets in an authentic, contextualized activity. Repeated viewings outside class time increases the success of second language learning outcome (Lo, 2024; Muñoz et al., 2022). Practice made perfect.

The sequential use of L1 subtitles and L2 captions is evidence to be a promising strategy for vocabulary learning through audiovisual materials (Markham et al., 2001; Wi & Boers, 2023; Yuan & Tang, 2025). This sequential approach has shown to be more effective in vocabulary gains compared to merely watching the video three times with only L2 captions (Wi & Boers, 2023). The results suggested that starting with L1 captions improve content comprehension, which in turn supports better vocabulary learning in subsequent viewings. The sequential caption usage also supports the progressive reduction of caption support (from full to partial to none) which helps learners transition from reliance on captions to independent listening, building confidence and real-world listening skills (Wi & Boers, 2023).

Innovative caption-based activity such as asking second language learners to create L2 captions for a L2 video can improve listening, reading, writing, and cultural awareness (Biasini & Raffi, 2023; Talaván, 2019). Learners can be asked to create captions for a short video clip with targeted linguistic elements or create their own creative L2 captions using a short video clip without audio track (Biasini & Raffi, 2023).

4.4. Selecting caption platform and video types

Access to information for language learning has evolved from the mere use of digital devices to the integration of digital platforms, such as YouTube, Vimeo, social media, and streaming services like Netflix and Home Box Office (HBO), which offer rich pedagogical opportunities through authentic content and multimodal resources (Montero Perez, 2022). While viewing YouTube content with captions offers the advantage of being free and therefore more accessible, paid streaming platforms typically provide professionally produced captions that may result in more effective learning outcomes.

Within a digital multimodal teaching environment, captioned videos can be used on platforms where videos can be uploaded and played back, such as various learning management systems including Canva, Blackboard, MOOC, and Moodle. While watched online or offline, these videos enable learners to self-pace their progress according to individual needs using slider control. This sense of control can increase learners' motivation, engagement, and confidence (Bjork et al., 2013), making the learning environment safe and predictable. Past studies evidenced the importance of learner perspectives affecting the effectiveness of L2 captions, and thus the pedagogical value of them (Pattemore & Muñoz, 2020; Pattemore et al., 2024; To, 2024).

Content selection for captioned videos is crucial for learner engagement. Appropriate topics, reading speed, and linguistic complexity enhance learning experiences (Vanderplank, 1988). For beginner learners who often require differentiated support, selecting videos that are simple in their visual layout with slower dialogue speed can reduce potential cognitive overload. Authentic videos featuring rapid spoken discourse present challenges for advanced learners, the use of L2 captions serves as valuable support to facilitate comprehension. Thoughtful selection of video materials while maintaining learning goals can improve outcomes and efficiency in language learning.

4.5 Dynamic approach in language learning

Larsen-Freeman (1997) claimed that language should be viewed as a dynamic system that is "dynamic, complex, nonlinear, chaotic, unpredictable, sensitive to initial conditions, open, self-organizing, feedback sensitive, and adaptive" (p. 147). Based on this view, a dynamic approach is proposed for selecting appropriate caption types, taking into account the multiple interacting components present in second language learning through authentic videos with captions. The effectiveness of caption types in second language learning depends on the dynamic interaction among learner characteristics, learning context, and instructional objectives. Learner characteristics refer to learners' language proficiency levels and individual differences; learning context refers to the environment and conditions in which language learning occurs, including the instructional setting, modality, sociocultural factors, availability of technological tools, task types, and exposure duration; and the instructional objectives refer to the intended learning outcome set by educators.

For instance, an advanced learner participating in an online academic English course may benefit from L2 captions when engaging with complex lectures, as these captions provide lexical and syntactic support without imposing excessive cognitive load, thereby promoting deeper processing of authentic input. In contrast, beginner learners in a face-to-face classroom with teacher support may achieve better comprehension and retention through L1 subtitles, which provide immediate semantic support, reduce cognitive demand, and allow learners to focus on meaning rather than processing unfamiliar forms. These examples illustrate that the strategic use of caption types should not be applied uniformly; rather, they should be adapted to align with the learner's proficiency, learning context, and the specific learning objectives, consistent with principles of cognitive load theory and the second language acquisition theories.

5. Pedagogical considerations and limitations

Effective use of captioned videos in language learning requires careful consideration of multiple pedagogical factors to optimize learning outcomes.

5.1. Accessibility and auto-caption accuracy

The traditional concept of captions for deaf or hard-of-hearing users has been challenged, so that different types of on-screen text can be accessible to and benefit all audiences regardless of their different abilities (Greco, 2019). Accessibility can also be improved by creating auto-captioning using automatic speech recognition technology on digital platforms.

However, auto-captions struggle with problems such as accuracy, speed, and text segmentation which divides text into meaningful chunks. A recent study showed that AI-generated captions have an average of 89.8% accuracy rate across four testing platforms including Kaltura (93.7%), Panopto (93.1%), Microsoft (87.7%), and Google (84.6%). The results demonstrated that these platforms cannot create auto-captions that meet the industry standard of at least 99% accuracy (Graham & Choo, 2022). Consequently, a high error rate in auto-captions affects the overall comprehension and learning experience. Thorough editing is necessary when using auto-captions in L2 videos for second language learning.

5.2. Learner cognitive load

In second language learning with captions, learners are often required to process information from multiple sources simultaneously, such as visual images, visual text, and sound, including background music and verbal speech. An increase in visual complexity in audiovisual inputs can impose extraneous load on learners, impeding learning (van der Zee et al., 2017). Research shows that L2 learners generally perform better in reading than listening (Danan, 2004; Garza, 1991), so L2 captions can aid advanced learners without adding cognitive load. However, for beginners, captions offer limited benefits because audiovisual input often exceeds their linguistic competence, making them less effective until stronger L2 skills are developed (Montero Perez, 2022).

Learners' attitude could affect their perception of cognitive load. For instance, learners with clear learning goals invest more effort in achieving learning outcomes. These learners are more likely to seek help with active engagement. Active learners may experience lower perceived cognitive load than their passive counterparts. Other individual factors may also affect learning, they include learners' beliefs, self-regulation, motivation, interest, engagement (Bjork et al., 2013), as well as learning strategies when given more control on learning pace (Costley et al., 2021).

5.3. Caption pacing and fatigue

Reading speed significantly influences learners' comprehension of video content, particularly for beginners. Authentic videos often feature rapid dialogue, which can overwhelm learners by exceeding their reading speed and cognitive processing capacity. For advanced learners, edited or reduced captions are recommended to lower cognitive load, while visually enhanced captions can also be used to focus on key language features. Fast-paced and auto-captioned videos should generally be avoided for beginner learners.

Captions that replicate natural speech rates can make it difficult for learners to follow both audio and text simultaneously. To address this, strategies such as reducing playback speed, segmenting captions into shorter chunks, and providing pre-viewing vocabulary support can improve comprehension and reduce cognitive load.

A virtual reality (VR) study showed that prolonged caption exposure can cause fatigue and lead to decreased engagement and increased stress, especially when users must maintain prolonged eye gaze to follow captions, a finding with clear implications for language learning contexts among language learners in general (Ubur et al., 2024).

5.4. Limitations of short-form videos

Recent trends in caption use on social media differ significantly from traditional practices, as the rise of online video platforms and social media (e.g., Meta, Instagram) has transformed how text is presented in digital formats such as YouTube Shorts, Instagram Reels, and Stories. On-screen texts on these digital platforms can be challenging for second language learners to process in an authentic environment because their display time is usually very short. This difficulty is further compounded when viewing on smartphones, where smaller screen sizes and disproportion caption size limit readability and increase cognitive load (Hwang et al., 2018).

These social media captions are unconventional and mostly creative and do not follow professional subtitling guidelines for optimal readability. Often their main purpose is content-oriented since these captions may be used to create an algorithm for online exposure to increase views and viewer engagements with financial benefits (Xing et al., 2020).

One of the common features of social media captions is that they are very fast, sometimes intentionally, so that their audience needs to re-watch their content to generate more views, but other times the creators may not be aware of the best practices. Kruger et al. (2022) found that viewers skipped 43% of words when the subtitle speed increased to 28 character per second, leading to shallow processing of information, resulting in reading errors and incomplete reading while unable to re-process unfamiliar content, reducing overall integration of the on-screen text and video.

The decision to use short videos as part of the language learning instruction must be carefully considered. Analyzing real-world data on language learning across digital platforms can shed light on caption-reading behaviors, including how fansub captions differ from professional ones and how various caption types influence effectiveness.

6. Future directions

Future research on captioning in educational contexts may benefit from exploring the potential of adaptive captioning. Captioning features such as pacing, lexical complexity, and bilingual pairing matched to individual learner profiles hold particular promise, particularly as multimodal learning has cemented itself as an essential learning environment since the COVID-19 pandemic. These features have the ability to enhance learner engagement, accessibility, and equity in learner-centered digital education. Undoubtedly, design choices regarding segmentation, font readability, and spatial placements can have cognitive implications and concerns; poorly calibrated captions may inadvertently increase extraneous load rather than reduce it, underscoring the need for authentic real-world studies beyond controlled laboratory settings.

Conducting further research in teachers' training and pedagogical opportunities is necessary. Effectiveness of captioning is based upon educators' training, attitudes, and instructional abilities. Without adequate teacher preparation, well-designed technological tools may still falter. Investigating how educators interpret, adapt, and integrate captioning into their practice may be important, as this could help identify the conditions under which captioning is more likely to enhance engagement rather than distract from it.

Finally, captioning in heritage or multilingual education contexts offers unique opportunities for research. Rather than treating captions as auxiliary aids, further research may explore how captions can function as support for heritage language maintenance, cross-linguistic transfer, and community-based learning. Examining the use of captions in bilingual classrooms, community media, and intergenerational programs may indicate how captions may contribute to equitable and culturally sustaining pedagogies. Future research can move beyond viewing captions as static add-ons by exploring their potential as dynamic, learner-centered tools. Integral to inclusive second language learning, the future of captions in enhancing second language learning is promising.

7. Conclusion

This chapter explores the strategic use of caption types through the lens of second language acquisition theories, positioning captions as a pedagogically flexible, evidence-based tool within digital multimodal environments. Drawing on dual coding theory and cognitive load theory, this chapter demonstrates how different caption types (e.g., L2 captions, L1 subtitles, and bilingual subtitles) can support language comprehension, vocabulary development, and content retention in distinct ways, depending on learner proficiency and instructional goals. Enhanced caption forms such as keyword, highlighted, and glossed captions further facilitate learning by directing attention to salient linguistic features and reducing cognitive overload. The proposed dynamic approach emphasizes the interaction among learner characteristics, learning contexts, and instructional objectives, offering a framework for more personalized and effective caption use. As second language learning continues to evolve alongside technological advancements, future research could focus on personalized and adaptive systems to suit individual learning needs, integration with advanced technologies, and long-term impact studies.

Acknowledgments

The authors would like to express their sincere gratitude to Ponn P. Mahayosnand, MPH, of the National Coalition of Independent Scholars, for her invaluable guidance in developing this manuscript. Her constructive feedback and insightful suggestions greatly enhanced the quality of this work. The first author also acknowledges Macquarie University for scholarship support and Professors Jan-Louis Kruger and Stephen Doherty for their guidance during the PhD research, from which part of this work originates.

References

Aldukhayel, D. (2021). The effects of captions on L2 learners' comprehension of vlogs. *Language Learning & Technology, 25*(2), 178–191. http://hdl.handle.net/10125/73439

Baddeley, A. (1992). Working memory. *Science, 255*(5044), 556–559. https://doi.org/10.1126/science.1736359

Bajorek, J. P. (2017). L2 pronunciation tools: The unrealized potential of prominent computer-assisted language learning software. *Issues and Trends in Learning Technologies, 5*(1). https://doi.org/10.2458/azu_itet_v5i1_bajorek

Bao, W. (2020). COVID-19 and online teaching in higher education: A case study of Peking University. *Human Behavior and Emerging Technologies, 2*(2), 113–115. https://doi.org/10.1002/hbe2.191

Biasini, R., & Raffi, F. (2023). For an interdisciplinary approach in language learning: Exploring the use of subtitling in the Italian language classroom. In A. Regagliolo (Ed.), *Italian as a foreign language: Teaching and acquisition in higher education* (pp. 233–250). Vernon Press. https://www.vernonpress.com/book/1758

Bjork, R. A., Dunlosky, J., & Kornell, N. (2013). Self-regulated learning: Beliefs, techniques, and illusions. *Annual Review of Psychology, 64,* 417–444. https://doi.org/10.1146/annurev-psych-113011-143823

Chan, W. S., Kruger J.-L., & Doherty, S. (2022). An investigation of subtitles as learning support in university education. *The Journal of Specialised Translation, 38,* 155–179. https://doi.org/10.26034/cm.jostrans.2022.087

Costley, J., Fanguy, M., Lange, C., & Baldwin, M. (2021). The effects of video lecture viewing strategies on cognitive load. *Journal of Computing in Higher Education, 33,* 19–38. https://doi.org/10.1007/s12528-020-09254-y

Danan, M. (2004). Captioning and subtitling: Undervalued language learning strategies. *Meta, 49*(1), 67–77. https://doi.org/10.7202/009021ar

Ellis, R. (2000). Task-based research and language pedagogy. *Language Teaching Research, 4,* 193–220. https://doi.org/10.1177/136216880000400302

Garza, T. J. (1991). Evaluating the use of captioned video materials in advanced foreign language learning. *Foreign Language Annals, 24*(3), 239–258. https://doi.org/10.1111/j.1944-9720.1991.tb00469.x

Gass, S., Winke, P., Isbell, D., & Ahn, J. (2019). How captions help people learn languages: A working-memory, eye-tracking study. *Language Learning & Technology, 23,* 84–104.

Graham, R., & Choo, J. (2022). Preliminary research on AI-generated caption accuracy rate by platforms and variables. *Journal on Technology and Persons with Disabilities, 10,* 33–53. http://hdl.handle.net/10211.3/223464

Greco, G. M. (2019). Accessibility studies: Abuses, misuses and the method of poietic design. *Lecture Notes in Computer Science (including Subseries Lecture Notes in Artificial Intelligence and Lecture Notes in Bioinformatics), 11786,* 15–27. https://doi.org/10.1007/978-3-030-30033-3_2

Hsieh, Y. (2019). Effects of video captioning on EFL vocabulary learning and listening comprehension. *Computer Assisted Language Learning, 33,* 567–589. https://doi.org/10.1080/09588221.2019.1577898

Hwang, G. J., Hsu, T. C., & Hsieh, Y. H. (2018). Impacts of different smartphone caption/subtitle mechanisms on English listening performance and perceptions of students with different learning styles. *International Journal of Human–Computer Interaction, 35*(4–5), 333–344. https://doi.org/10.1080/10447318.2018.1543091

Krashen, S. (1982). *Principles and practice in second language acquisition.* Pergamon Press.

Kruger, J., Wisniewska, N., & Liao, S. (2022). Why subtitle speed matters: Evidence from word skipping and rereading. *Applied Psycholinguistics, 43*(1), 211–236. https://doi.org/10.1017/S0142716421000503

Kurokawa, S., Hein, A., & Uchihara, T. (2024). Incidental vocabulary acquisition through captioned viewing: A meta-analysis. *Language Learning.* https://doi.org/10.1111/lang.12697

Larsen-Freeman, D. (1997). Chaos/Complexity science and second language acquisition. *Applied Linguistics, 18*(2), 141–165. https://doi.org/10.1093/applin/18.2.141

Leahy, W., & Sweller, J. (2011). Cognitive load theory, modality of presentation and the transient information effect. *Applied Cognitive Psychology, 25*(6), 943–951. https://doi.org/10.1002/acp.1787

Lee, M., & Révész, A. (2020). Promoting grammatical development through captions and textual enhancement in multimodal input-based tasks. *Studies in Second Language Acquisition, 42*(3), 625–651. https://doi.org/10.1017/S0272263120000108

Liao, S., Kruger, J. L., & Doherty, S. (2020). The impact of monolingual and bilingual subtitles on visual attention, cognitive load, and comprehension. *The Journal of Specialised Translation, 33,* 70–98. https://doi.org/10.26034/cm.jostrans.2020.549

Lo, S. (2024). Vocabulary learning through viewing dual-subtitled videos: Immediate repetition versus spaced repetition as an enhancement strategy. *ReCALL, 36*(2), 152–167. https://doi.org/10.1017/S0958344024000053

Markham, P. (1999). Captioned videotapes and second-language listening word recognition. *Foreign Language Annals, 32*(3), 321–328. https://doi.org/10.1111/j.1944-9720.1999.tb01344.x

Markham, P., Peter, L. A., & McCarthy, T. J. (2001). The effects of native language vs. target language captions on foreign language students' DVD video comprehension. *Foreign Language Annals, 34*(5), 439–445. https://doi.org/10.1111/j.1944-9720.2001.tb02083.x

Montero Perez, M. (2019). Pre-learning vocabulary before viewing captioned video: an eye-tracking study. *The Language Learning Journal, 49*(4), 460–478. https://doi.org/10.1080/09571736.2019.1638623

Montero Perez, M. (2022). Second or foreign language learning through watching audio-visual input and the role of on-screen text. *Language Teaching, 55,* 163–192. https://doi.org/10.1017/S0261444821000501

Montero Perez, M., Peters, E., & Desmet, P. (2018). Vocabulary learning through viewing video: The effect of two enhancement techniques. *Computer Assisted Language Learning, 31*(1–2), 1–26. https://doi.org/10.1080/09588221.2017.1375960

Montero Perez, M., Van Den Noortgate, W., & Desmet, P. (2013). Captioned video for L2 listening and vocabulary learning: A meta-analysis. *System, 41*(3), 720–739. https://doi.org/10.1016/j.system.2013.07.013

Muñoz, C., Pattemore, A., & Avello, D. (2022) Exploring repeated captioning viewing as a way to promote vocabulary learning: Time lag between repetitions and learner factors. *Computer Assisted Language Learning, 37*(7), 1744–1770. https://doi.org/10.1080/09588221.2022.2113898

Muñoz, C., Pujadas, G., & Pattemore, A. (2023). Audio-visual input for learning L2 vocabulary and grammatical constructions. *Second Language Research, 39*(1), 13–37. https://doi.org/10.1177/02676583211015797

Paivio, A. (1991). Dual coding theory: Retrospect and current status. *Canadian Journal of Psychology, 45*(3), 255–287. https://doi.org/10.1037/h0084295

Pattemore, A., & Muñoz, C. (2020). Learning L2 constructions from captioned audio-visual exposure: The effect of learner related factors. *System, 93*, Article 102303. https://doi.org/10.1016/j.system.2020.102303

Pattemore, A., Suárez, M., & Muñoz, C. (2024). Perceptions of learning from audiovisual input and changes in L2 viewing preferences: The roles of on-screen text and proficiency. *ReCALL, 36*(2), 135–151. https://doi.org/10.1017/S0958344024000065

Pujadas, G., & Muñoz, C. (2024). When to switch captions off? Exploring the effects of L2 proficiency and vocabulary knowledge on comprehension of captioned and uncaptioned TV. *Studies in Second Language Learning and Teaching, 14*(3), 545–570. https://doi.org/10.14746/ssllt.38036

Roshne, V., Balaji, S. R., Sowndharyan, V. R. S., & Vivek, C. (2024). *Empowering content creation using artificial intelligence - The role of caption writing: An overview* [Paper presentation]. International Conference on Science Technology Engineering and Management (ICSTEM), India. https://doi.org/10.1109/ICSTEM61137.2024.10560691

Schmidt, R. (1990). The role of consciousness in second language learning. *Applied Linguistics, 11*(2), 129–158. https://doi.org/10.1093/applin/11.2.129

Swain, M. (1985). Communicative competence: Some roles of comprehensible input and comprehensible output in its development. In S. Gass & C. Madden (Eds.), *Input in Second Language Acquisition* (pp. 235–253). Newbury House. https://archive.org/details/inputinsecondlan0000unse

Sweller, J. (2011). Cognitive load theory. *Psychology of Learning and Motivation, 55*, 37–76. https://doi.org/10.1016/B978-0-12-387691-1.00002-8

Sweller, J. (2024). Cognitive load theory and individual differences. *Learning and Individual Differences, 110*, 102423. https://doi.org/10.1016/j.lindif.2024.102423

Talaván, N. (2019). Creative audiovisual translation applied to foreign language education: A preliminary approach. *Journal of Audiovisual Translation, 2*(1), 53–74.

Teng, M. F. (2020). Vocabulary learning through videos: captions, advance-organizer strategy, and their combination. *Computer Assisted Language Learning, 35*, 1–33. https://doi.org/10.1080/09588221.2020.1720253

To, C. (2024). Are subtitles useful for language learners? *Journal of Language Teaching, 4*(2), 1–6. https://doi.org/10.54475/jlt.2024.006

Ubur, S. D., Etori, N. A., Ghasemi, S., King, K., Gračanin, D., & Gini, M. (2024). EasyCaption: Investigating the impact of prolonged exposure to captioning on VR HMD on general population. In M. Antona, and C. Stephanidis (Eds), *Universal access in human-computer interaction. HCII 2024. Lecture notes in computer science, 14697*. Springer. https://doi.org/10.1007/978-3-031-60881-0_24

Vanderplank, R. (1988). The value of teletext sub-titles in language learning. *ELT Journal, 42*(4), 272–281. https://doi.org/10.1093/elt/42.4.272

Van Der Zee, T., Admiraal, W., Paas, F., Saab, N., & Giesbers, B. (2017). Effects of subtitles, complexity, and language proficiency on learning from online education videos. *Journal of Media Psychology Theories Methods and Applications, 29*, 18–30. https://doi.org/10.1027/1864-1105/a000208

Wang, A., & Pellicer-Sánchez, A. (2022). Incidental vocabulary learning from bilingual subtitled viewing: An eye-tracking study. *Language Learning, 72*(3), 765–805. https://doi.org/10.1111/lang.12495

Wei, R., & Fan, L. (2022). On-screen texts in audiovisual input for L2 vocabulary learning: A review. *Frontiers in Psychology, 13*, 904523. https://doi.org/10.3389/fpsyg.2022.904523

Wi, I., & Boers, F. (2023). Sequential use of L1 and L2 captions: Exploring the benefits for vocabulary acquisition. *TESOL Quarterly, 58*, 511–521. https://doi.org/10.1002/tesq.3243.

Xing, F., Poria, S., Cambria, E., & Welsch, R. (2020). Social media marketing and financial forecasting. *Information Processing & Management, 57*(5), 102314. https://doi.org/10.1016/j.ipm.2020.102314.

Yuan, X., & Tang, X. (2025). Effects of the sequential use of L1 and bilingual subtitles on incidental English vocabulary learning: A cognitive load perspective. *The British Journal of Educational Psychology*. https://doi.org/10.1111/bjep.12740.

Chapter 11

Adaptive learning in second language teaching: insights for practice

Anne Prucha

University of Central Florida

Kacie Tartt

University of Central Florida

Abstract

This chapter discusses the integration of personalized adaptive learning (PAL) and open educational resources (OER) in the redesign of the first-year Spanish course sequence at the University of Central Florida. Leveraging the PAL platform Realizeit, which dynamically adjusts learning paths based on individual student performance, the redesign addressed disparities in prior knowledge and improved access to instructional materials. The replacement of commercial textbooks with OER further supported equitable learning environments. Quantitative data from the redesigned courses indicate significant gains in student engagement, retention, and academic achievement, with particularly notable improvements in ABC grade rates among minoritized student populations. These outcomes suggest that the combined use of PAL and OER fosters more inclusive, student-centered pedagogies. Our instructional team observed immediate benefits from the redesign and now seeks to disseminate these findings to encourage critical reflection on current practices and to support continued innovation in language education.

Keywords: Adaptive learning, Open educational resources, Inclusive pedagogy, Course redesign, Student engagement and achievement

* * *

1. Introduction: a narrative of innovation in language teaching and learning

Foreign language instruction in higher education has long contended with the challenges of addressing learner diversity, ensuring equitable access to materials, and sustaining student engagement, particularly in contexts with limited exposure to the target language. In this chapter, we share how our instructional team at the University of Central Florida (UCF) reimagined our first-year Spanish sequence by integrating personalized adaptive learning (PAL) and transitioning from publisher courseware to open educational resources (OER), all through the lens of second language acquisition (SLA) theory.

UCF is a large metropolitan university that enrolls nearly 70,000 students per year. It is also a Hispanic serving institution (HSI), a federal designation assigned by the U.S. Department of Education to colleges and universities whose full-time undergraduate student body is 25% or more Hispanic (U.S. Department of Education, n.d.). Hispanic students make up 30.2% of the total student body at UCF, with 51.5% of the student body being students of color. These statistics underscore the immense diversity at the team's institution.

The Elementary Spanish Language and Civilization course sequence at UCF enrolls approximately 5,000 students per year across four semesters. These courses introduce students to Hispanic cultures through listening, speaking, reading, and writing. The first course is designed for students with no prior knowledge of Spanish and serves as a prerequisite for the second. Between Fall 2023 and Summer 2024, 22% of students were enrolled in PAL/OER sections—a proportion that has grown steadily since 2018.

SLA research emphasizes individual differences, form-focused instruction, noticing, and learner autonomy (Ellis, 2005; Long, 1991; Schmidt, 1990), principles equally crucial in foreign language (FL) classrooms, where authentic input is limited. Our narrative illustrates how PAL and OER operationalize these principles, aligning with communicative and task-based teaching trends (Larsen-Freeman & Anderson, 2011; VanPatten, 2015).

This chapter offers a reflective account of our journey: the motivations behind our redesign, the pedagogical decisions we made, the challenges we encountered, and the insights we gained. By weaving together theory, practice, and lived experience, we aim to show how adaptive technologies, and open resources can serve as pedagogical tools, and not just technological solutions, for creating more inclusive, responsive, and effective language learning environments. In doing so, we hope to contribute to a broader conversation about innovation in foreign language education. Our story is one of transformation—not only of course design, but of instructional mindset—and we offer it as an invitation for others to reflect, adapt, and imagine new possibilities in their own teaching contexts.

2. Challenges of traditional course delivery

The problems the instructional team aimed to solve by redesigning courses with PAL and OER were multifaceted. Year after year, students who enroll in first-year Spanish courses at UCF have prior knowledge of Spanish, even though the first course in the sequence assumes no knowledge of the language. This often stems from students having studied Spanish in middle or high school, with a typical gap of one to three years between their last class and the start of their studies at UCF. Additionally, due to the cultural makeup of the State of Florida, where nearly 30% of the population is Hispanic, and UCF's HSI status, many students have family members who speak Spanish. As a result, it is not uncommon for students to be from heritage Spanish-speaking backgrounds or from areas or neighborhoods where Spanish speakers are prevalent (Florida Office of Economic and Demographic Research, 2024).

Additionally, many students are unable to access course materials due to economic factors; for example, late financial aid disbursements or lack of funds to pay for traditional textbooks and publisher courseware. Indeed, in many cases students forgo purchasing textbooks and other course materials so that they have funds for other necessities (Scholarly Publishing and Academic Resources Coalition [SPARC], 2023).

3. Course design framework

3.1. Personalized adaptive learning

Adaptive learning (termed personalized adaptive learning, or PAL at UCF), is a data-driven instructional technique that customizes learning experiences to meet each student's individual needs. Adaptive learning software tracks data on student progress, engagement, and performance to create personalized learning experiences. Adaptive learning produces a more efficient, engaging, and unique learning experience, the idea being that this will lead to better outcomes for students and more effective use of educational resources. Dziuban, Moskal, Parker, and Campbell (2018) recognize that "those technological developments may alter the teaching and learning process, but adaptive learning also modifies the most critical factor—time" (p. 2). Personalizing the learning process through adaptive tools allows students to spend their time on what they need to learn and not what they already know.

The Realizeit software platform assesses students' interactions with course content to determine their current knowledge, identifies what they need to learn next, and creates an appropriate learning pathway. As students advance through the content, the system continuously updates and refines their pathways based on ongoing participation, selecting the most suitable content and tailoring lessons to their learning preferences. This approach allows students to focus on what they need to learn by recognizing their existing knowledge and addressing their individual needs. What is more, Realizeit offered the team comprehensive assessment tools, integration with our existing learning management system, scalability, support for diverse learning needs, flexible implementation, and continuous improvement of our delivery methods. The platform also offers robust learning analytics (Siemens, 2013). Having immediate access to data analytics contributes to the positive impact that PAL has on student learning and success. (Khor & Mutthulakshmi, 2024).

This adaptive system closely aligns with the principles of form-focused instruction (FFI), a key component of SLA pedagogy that emphasizes drawing learners' attention to linguistic form within meaningful communicative contexts. While FFI has traditionally been explored in immersion or second language environments, its principles are equally applicable to foreign language classrooms, where structured opportunities for noticing and feedback are especially valuable due to limited exposure to the target language. As Long (1991) and Doughty and Williams (1998) argue, FFI is most effective when integrated into communicative tasks and supported by timely corrective feedback. Realizeit's ability to provide instant, individualized feedback supports this approach by helping learners notice and correct errors as they occur, an essential process for developing grammatical accuracy and fluency, regardless of the instructional setting. This is further supported by Nassaji and Fotos (2011), who emphasize the value of integrating grammar instruction into communicative contexts.

Incorporating Realizeit into the team's Spanish courses has transformed both the delivery of educational content and the way students engage with it. As part of the course redesign, Realizeit served as the primary medium for disseminating instructional material and facilitating vocabulary and grammar practice exercises. Its adaptive features customized the learning experience to accommodate the individual pace, prior knowledge, and preferences of each student as they progressed through the course units. This personalization ensured that students received immediate, targeted feedback and advanced through the material in a way that acknowledged their existing competencies while addressing areas for growth.

Furthermore, the platform's scaffolded learning pathways and optional remedial activities allow students to revisit and reinforce previously covered content. This cyclical engagement with form not only supports noticing and uptake, two key mechanisms in SLA, but also empowers learners to take ownership of their progress. By enabling students to self-direct their review and remediation, Realizeit fosters autonomous learning, another core goal of FFI-based pedagogy.

In this way, Realizeit does more than deliver content; it creates a responsive, form-sensitive learning environment that mirrors best practices in SLA. The integration of adaptive technology with FFI principles enhances both the efficacy and equity of instruction, ensuring that all learners have the opportunity to achieve course goals and develop robust linguistic competence.

3.2. Open educational resources

To enhance the team's adaptive approach and increase accessibility, we integrated OER into the course redesign. OER are defined by United Nations Educational, Scientific, and Cultural Organization (UNESCO) as learning, teaching, and research materials in any format and medium that reside in the public domain or are under copyright but released under an open license, permitting no-cost access, reuse, repurposing, adaptation, and redistribution by others (UNESCO, n.d.). OER can either belong to the public domain or be covered by more flexible intellectual property licenses. They can be modified, combined with other materials, expanded upon, translated, and then redistributed to suit various needs. OER come in many formats and can be used for myriad teaching and learning purposes, including course content like e-textbooks (or content in lieu of a textbook, as utilized in our course redesign), activities, and assessments. They are not restricted to any specific learning platform (Affordable Learning LOUISiana, n.d.).

OER play a pivotal role in contemporary education, offering educators a vast array of adaptable, localized, and free resources to enhance learning experiences for diverse student populations. We took advantage of these freely accessible supplementary materials and tailored them to complement the Realizeit platform. With the support of an instructional designer, we redesigned the courses by creating a detailed scope and sequence, curating relevant OER materials, and integrating new, self-authored content alongside existing resources. The redesign includes interactive exercises, multimedia content, and readings that enrich the learning experience and provide varied perspectives on the course material. By replacing costly textbooks and proprietary learning management systems, the instructional team has fostered a more personalized and engaging educational environment.

We used OER to gain greater freedom and flexibility, allowing the selection of relevant, engaging resources that resonated with students, thereby increasing their motivation and satisfaction. The ability to adapt and modify content enabled us to create a dynamic curriculum that reflects current cultural contexts (in our case, for example, comparing and contrasting Hispanic cultures with American and local cultures), engaging students more effectively than with traditional materials. This not only benefited student learning outcomes but also enhanced the team's job satisfaction as we witnessed the positive impact of these innovative teaching methods.

The team made changes to the course content as needed without relying on a publisher. This flexibility allowed us to incorporate appropriate, relevant, and engaging content, and to create and deliver meaningful practice and assessments. Students were encouraged to share their interests regarding course topics and themes, enabling us to tailor grammar and culture content to individual semesters and highlight areas of student interest we might not have previously considered. These practices directly impacted student success and engagement.

Both PAL and OER are integral to the future of modern higher education for several reasons: enhanced student engagement, data-driven insights, flexibility and accessibility, improved learning outcomes, support for faculty, cost-effectiveness, scalability, and preparation for future careers. This approach aligned with the team's personal goals for the course redesign, as well as the guidelines set out by the Course Redesign Initiative (CRI) administered by UCF's Pegasus Innovation Lab. Institutional support was vital, providing resource allocation, networking opportunities, expertise and guidance, administrative support, legitimacy and credibility, and training and development. This support allowed the team not only to improve our courses but also to pave the way forward in our discipline.

4. Addressing student diversity

One of the primary advantages of implementing PAL in Spanish courses is its ability to address the diverse needs of students. For this redesign we adopted the platform Realizeit, which utilizes adaptive learning technology that tailors the educational experience to each student's unique background, prior knowledge, and learning preferences. By assessing students' existing knowledge and skills, Realizeit creates personalized learning paths that guide students through the course at an appropriate pace, ensuring that they are neither overwhelmed nor under-challenged. This personalized approach helps bridge gaps in knowledge and provides enrichment where needed, thereby fostering a more inclusive and effective learning environment.

The adaptive approach is grounded in the principles of differentiated instruction, a pedagogical framework that has gained increasing traction in SLA. While our project centers on foreign language pedagogy, typically characterized by more limited exposure to the target language outside the classroom, it's important to recognize that the core tenets of differentiated instruction remain highly relevant. Differentiated instruction acknowledges that all language learners, whether acquiring a second language in an immersive environment or studying a foreign language in a more constrained setting, bring diverse linguistic repertoires, cognitive styles, cultural backgrounds, and affective needs to the classroom. As Lightbown and Spada (2013) and Ellis (2005) emphasize, effective language instruction must be responsive to these differences, offering multiple pathways to learning that accommodate individual variability. In this context, Realizeit functions not just as a delivery tool but as a dynamic instructional partner, enabling educators to implement differentiation at scale, even in foreign language classrooms where learner diversity is just as pronounced as in second language classrooms. As Tomlinson (2014) notes, effective language instruction must be flexible and responsive to learners' evolving needs, particularly in diverse classrooms.

Moreover, the platform's flexibility, enhanced by the integration of OER, allows instructors to select and adapt content that resonates with learners' interests, goals, and identities. This is particularly important in language learning, where motivation and personal relevance are closely tied to successful outcomes. The system's ability to provide real-time feedback and scaffolded support ensures that learners receive timely, targeted interventions, which are a hallmark of effective differentiated instruction.

Additionally, Realizeit provides educators with valuable analytics on student performance and engagement. These insights serve as a roadmap for instructors, helping them identify learning trends, predict potential challenges, and intervene proactively. In doing so, the platform supports a responsive, learner-centered environment where all students, regardless of their starting point, can achieve the course goals and realize their full potential in learning Spanish.

5. Technical and financial barriers

We had previously observed that many students faced challenges with online delivery modes and high textbook costs, negatively impacting retention and success rates (Colvard, Watson, & Park, 2018). Using Realizeit and OER allowed us to eliminate for students the financial burden of expensive textbooks, making course materials accessible at no or low cost. This accessibility not only alleviated financial stress but also contributed to improved course completion rates. Data analysis revealed a significant increase in the proportion of students achieving ABC grades after the adoption of PAL and OER, highlighting the platform's effectiveness in enhancing learning outcomes and reducing drop, fail, withdraw (DFW) rates among all students, and particularly among minoritized students. With this preliminary data, we felt confident that we were moving in the right direction.

The integration of Realizeit into the team's lower-level Spanish courses also addressed technical challenges associated with traditional publisher content and courseware platforms, which often prevented students from commencing the course in a timely manner. Students reported a smoother, more intuitive learning experience with immediate feedback and structured pathways. This approach minimized frustrations often encountered in online learning environments, thereby fostering a supportive and inclusive atmosphere conducive to learning. With reduced stress, we were able to focus more on supporting our students and delivering quality content, rather than troubleshooting technical problems as we had had to do prior to the redesign.

Students repeatedly praised the Realizeit platform for its affordability, ease of use, and ability to build confidence in language learning. One student noted, "I appreciated not needing to purchase a book..." (Anonymous student, course evaluation, December 2020) while another shared, "The course was very well-done and easy to handle as a beginner." (April 2020) The shift to OER reduced financial barriers by eliminating the need for costly textbooks, which traditionally hindered access to course materials. This affordability not only promoted equal access but also aligned with UCF's commitment to enhancing student success and retention.

Researchers at UCF have been conducting a Textbook Affordability Survey that includes students in the team's PAL/OER Spanish courses since Spring 2021. Nine percent of students in these courses have responded to this voluntary survey during this time frame. According to their findings, access to free course materials could directly influence how students spend funds; excluding additional education costs, students reported housing, transportation, and food as the top targets of additional spending (Figure 11.1). This ongoing research aims to explore how low or no-cost textbooks affect student behaviors and outcomes (Center for Distributed Learning, UCF, n.d.).

In our experience, students have not been stimulated or motivated by traditional publisher content. The reasons for this vary; they have used Spanish-language textbooks in middle and/or high school, the textbook content lacks relevance, or the courseware platforms pose technical challenges. When students encounter technical barriers to courseware it can lead to them falling behind. By eliminating reliance on publisher courseware systems this challenge has been eliminated, directly contributing to increased student engagement and success. Additionally, PAL technology and OER content have allowed us to design the courses to be more personal, more appealing, and more meaningful to students, infusing them with relevant cultural content and "real-world" knowledge. This knowledge is transferable to the post-graduation sphere, in careers, personal life and/or graduate school. On the personal side, PAL creates an individual learning path for students, allowing them to focus on the concepts they do not yet know or for which they need a stronger foundation.

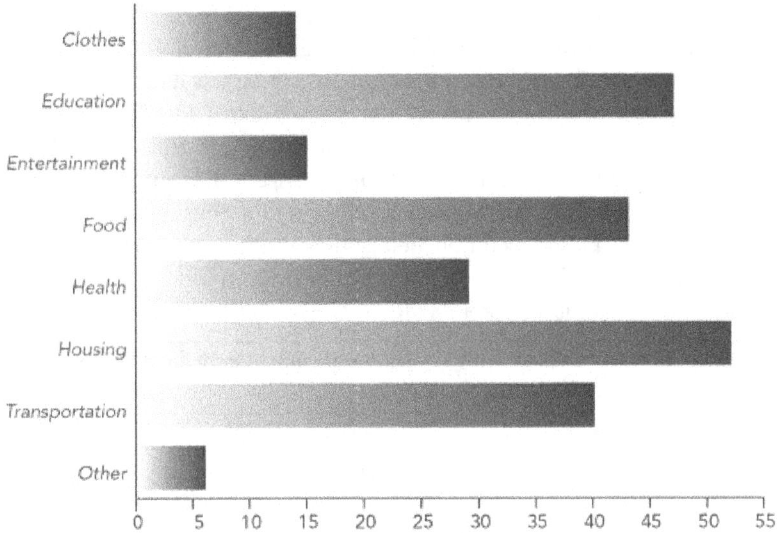

Figure 11.1. How students spent money saved on course materials

The results of the Textbook Affordability Survey (Figure 11.1) reflect the responses from students regarding how the cost of course materials directly impact their decision-making. The same survey also gauged students' textbook purchasing habits, focusing on the impact of textbook costs on students' academic decisions. Around 65–70% of students indicated that they did not purchase or that they delayed the purchase of textbooks due to cost. Additionally, many in this group shared textbooks, took fewer courses, or chose not to enroll in a course for the same reason. Overall, the data highlight the substantial influence of textbook costs on students who acknowledge financial barriers. Additionally, students in the PAL/OER courses reported that the PAL/OER content offered numerous academic and personal benefits through its course materials and practice activities.

6. Criteria for choosing Realizeit

The adoption of Realizeit as the digital learning tool for the Elementary Spanish Language and Civilization courses at UCF was driven by several key criteria and concerns. Below is a detailed breakdown of the considerations upon which this choice was based.

6.1. Cost effectiveness for students

Realizeit was chosen primarily because it provided free or low-cost access to students. In fully online classes through summer 2024, students did not have to pay for access, while hybrid classes required a nominal fee, approximately $35. This is aligned with UCF's Affordable Instructional Materials (AIM) Initiative (https://digitallearning.ucf.edu/ilab/aim/), which targets a reduction in costs for students through the inclusion of content like OER. The university's commitment to affordability and accessibility through initiatives like AIM underscores the strategic decision to adopt Realizeit. AIM supports faculty in adopting low-cost or no-cost materials, allowing for continuity with broader educational goals. This is accomplished principally through the initiative's four main pillars: First Day, Open Educational Resources, Library-Sourced Materials, and Affordability Counts (Table 11.1).

Pillars of Affordable Instructional Materials (AIM)	Description of Individual Pillars
First Day	Implementation of an inclusive access program through the UCF Bookstore in collaboration with publishers and other for-profit entities offering discounts on course content and/or courseware solutions
Open Educational Resources	Free teaching and learning materials that can be used, adapted, and shared with few or no restrictions.
Library-Sourced Materials	UCF libraries helping faculty find library eBooks that can fully replace course texts, focusing on titles with unlimited user access and favorable Digital Rights Management (DRM) terms
Affordability Counts	Program led by Florida International University (FIU) that recognizes faculty who reduce textbook costs by adopting affordable course materials across Florida colleges and universities.

Table 11.1. Description of four main pillars of the AIM Initiative

6.2. Ability to address diverse student backgrounds and levels

The elementary Spanish courses at UCF attract students with a wide range of language backgrounds and proficiencies. Realizeit's adaptive learning (PAL) platform offered an individualized learning experience, allowing students to progress at their own pace and focus on areas where they needed improvement, accommodating both beginners and those with prior Spanish knowledge.

6.3. Understanding our students

Student backgrounds, including those of minoritized and neurodiverse learners, played a central role in our course redesign. Initially, our focus was on students' prior knowledge, but deeper research revealed that our innovations had broader impacts, especially on how external responsibilities and pressures (Tinto, 2017) shaped students' learning experiences. With data collected from our courses, the instructional team now customizes content to better support these populations and maintain affordability across classes.

6.4. Key features of Realizeit

Realizeit offers numerous features and characteristics. Two that contributed to learner success were personalized learning paths and robust analytics and monitoring tools. The platform identifies what individuals know and don't know, then adapts course content based on their existing knowledge, skills, and learning needs, ensuring engagement with material that aligns with their level. This approach helps direct attention toward unlearned concepts and knowledge gaps, rather than revisiting areas of established proficiency.

Realizeit's robust analytics tools also provide information on student performance; for example, assigned work not yet begun or not completed by the due date, time spent on review and practice, and even emotional reactions (with the use of emojis), among others. Additionally, by employing a tool that allowed the team to use OER, we were able to integrate relevant and engaging content that resonated with students, enhancing their learning experience.

6.5. Comparison with other tools

While specific comparisons with other adaptive tools, like Acrobatiq and publisher courseware systems, will not be detailed here, the decision to adopt Realizeit was influenced by its comprehensive features, cost-effectiveness, and the support it offered for diverse student needs. Furthermore, in the spirit of accessibility, Realizeit was the tool where we had the most support as instructors, not only from UCF, but from the company itself. At UCF, adaptive learning has been primarily used in STEM courses. Recognizing its potential to enhance learner success, we became early adopters of the technology in the Humanities, specifically in language instruction. This pioneering effort introduced a learning curve across the board, affecting instructors, course developers, and support teams alike.

7. Enhancing student motivation and engagement

Our experience has shown that students experience fatigue with traditional publisher content; it often fails to resonate with them, leading to disengagement and reduced motivation. We recognized this challenge and have even experienced it from the instructor perspective, another factor in the team's decision to redesign our courses using PAL and OER. By incorporating real-world, relevant content and adaptive learning paths, learning experiences were customized to individual student interests and needs and the result was a more engaging and meaningful learning environment, which significantly boosted student satisfaction and participation. This increase in motivation aligns with Dörnyei's (2009) L2 Motivational Self System, which posits that learners are more engaged when instructional content is personally meaningful and aligned with their ideal language self.

8. Measuring the impact of Realizeit

To assess the impact of Realizeit, we collaborated with a colleague to conduct an survey, comparing success rates before and after implementing PAL and OER in our courses. We specifically measured the percentage of students achieving ABC grades and the rates of students receiving DFW grades. Our analysis demonstrated a significant improvement in student success rates, particularly among minoritized students, following the adoption of Realizeit. Additionally, we observed positive trends in student retention and overall satisfaction. This research included minoritized students enrolled from Spring 2014 through Fall 2023. According to § 365(3) of the HEA (3) (20 U.S.C. § 1067k(3)), "minority" is defined as American Indian, Alaska Native, Black, Hispanic, and Pacific Islander (UCF, n.d.).

Of the minoritized students enrolled in the Spanish I course during that period, 473 had taken the course—129 prior to PAL/OER and 344 in the PAL/OER sections. An independent samples t-test data allowed us to compare the mean percentage of students who succeeded (ABC grade) across the two groups, before PAL/OER and with PAL/OER. The analysis indicated that minoritized students in PAL sections had a significantly higher proportion of ABC grades (M=0.904; SD = 0.295) compared to those in the courses prior to PAL/OER (M=0.721; SD = 0.295), $t(170.845) = 4.287$, $p < .001$.

Similarly, of the minoritized students enrolled in the Spanish II course, 611 had taken the course—230 prior to PAL/OER and 381 in the PAL/OER sections. The independent t-test analysis revealed that minoritized students in PAL/OER sections had a significantly higher proportion of ABC grades (M=0.916; SD = 0.278) compared to those in the course before PAL/OER (M=0.700; SD = 0.459), $t(331.54) = 6.46$, $p < .001$.

Figures 11.2 and 11.3 show these proportions converted to percentages. In both Spanish I and Spanish II, the introduction of PAL/OER significantly increased the number of students who successfully completed each course with an ABC grade and significantly reduced the percentage of students who did not succeed (DFW).

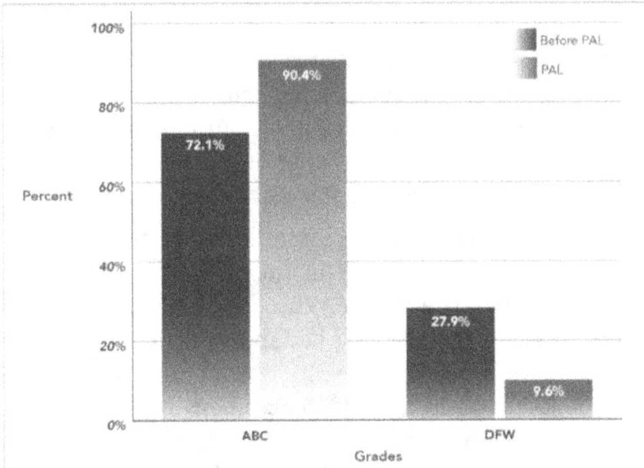

Figure 11.2. Impact of personalized adaptive learning (PAL) on grade outcomes for minoritized students in Spanish I: Comparison of ABC grades vs. DFW rates before and after PAL implementation

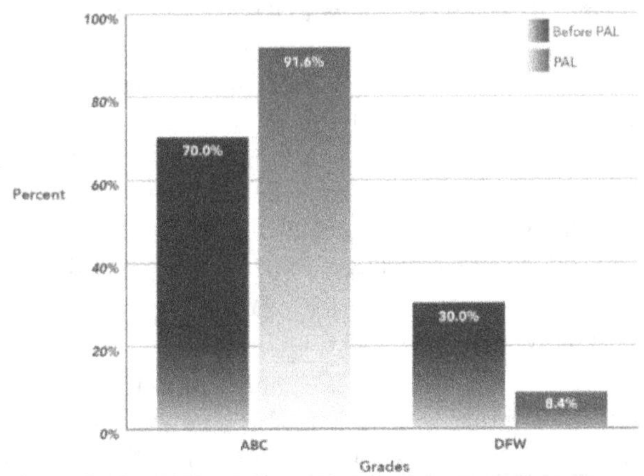

Figure 11.3. Impact of personalized adaptive learning (PAL) on grade outcomes for minoritized students in Spanish II: Comparison of ABC grades vs. DFW rates before and after PAL implementation

The overall response from students to adaptive learning has been overwhelmingly positive for several reasons. Clear expectations regarding the use of Realizeit and OER were posted in the welcome module of each course. The grading procedure within the platform was explained in detail, and students were instructed to repeat the material, with the learning content and practice activities always available. Students reported feeling more empowered and engaged with their learning, which contributed to increased confidence in their language skills.

Special care was taken to provide a large enough bank of material to prevent content repetition. Students were never "locked out" of the content or practice activities. The goal was to inform students about the "how" and "why" of the tool's importance and benefits before they began using it. This way, they saw the value of this structure, especially since it was no or low cost.

Negative feedback generally stemmed from students who had begun the course sequence—Spanish I with an instructor who had used the traditional textbook and publisher courseware. This led to feelings of unfamiliarity, compounded by the fact that students had spent nearly $300 on materials they ended up using for only one semester. It's worth noting that the instructional team informs students about the differences between tour course materials and the traditional textbook and publisher courseware used in other class sections before the semester starts, but some still slip through the cracks.

In recent years, educational technology (El-Sabagh, 2021) has revolutionized course design and student engagement, significantly enhancing the team's online Spanish language courses. As previously mentioned, the instructional team have collaborated with colleagues to redesign additional Spanish courses as well, intermediate and advanced. By integrating PAL and OER in general education courses, as well as courses required for the Spanish major and minor, we've aimed to and been successful in improving student learning outcomes (Eau, Judah, & Shahid, 2019) and making language education more accessible and engaging for a diverse student body. Beyond improving our pedagogical approach and increasing student engagement and success, our initiative seeks to impact Spanish language education at UCF and beyond. By showcasing the effectiveness of PAL and OER in introductory courses, we aim to inspire more students to pursue Spanish majors or minors.

Our ongoing course redesign and integration of PAL and OER exemplify a forward-thinking approach to language education. By embracing technological innovation and collaborative design practices, we have improved the quality and accessibility of our first-year Spanish courses, setting a precedent for future advancements in educational practices. Looking ahead, continued adaptation and refinement of these methodologies promise continued investment in students and educators, ensuring a more personalized and effective learning experience.

Institutions can adopt similar initiatives to innovate and optimize educational experiences, shaping a more inclusive and effective educational landscape for all learners. Faculty members can take pride in their role in this achievement, as their commitment to creating an inclusive and supportive learning environment directly contributes to increased student engagement and retention.

9. Student feedback and impact

Feedback from students regarding Realizeit has been overwhelmingly positive. Students appreciated the platform's user-friendly interface, personalized learning paths, and the availability of resources at no or low cost. They emphasized the value of immediate feedback and the ability to revisit materials as needed, which contributed to their understanding and retention of course content. A student enrolled in SPN II, succinctly expressed their perspective:

> I believe that using the Realizeit system is the most effective way to learn a foreign language. I only had to learn concepts that I was not the greatest at or never learned. Which I think is the most efficient form of learning (Anonymous student, course evaluation, December 2020).

Another student, enrolled in Spanish I, said, "As someone with very little confidence in language learning abilities, the course was very well done and easy to handle as a beginner. Not having to purchase a textbook was wonderful" (Anonymous student, course evaluation, December 2021).

This type of feedback not only validated our efforts but also motivated us to continue exploring and implementing new strategies and methodologies to enhance language education. Our course redesign employing Realizeit and OER demonstrates the transformative potential of PAL and OER in higher education. By leveraging adaptive learning technologies and open resources, we were able to successfully address longstanding challenges related to student engagement, retention, and financial accessibility. The positive outcomes observed emphasize the importance of ongoing course redesign and innovation in meeting the evolving needs of students in today's digital-native era. As higher education continues to progress, our approach, supported by our university community, serves as a model for institutions seeking to enhance learning outcomes and support student success through personalized, adaptive learning environments.

The adoption of Realizeit and OER represents a paradigm shift towards more inclusive, effective, and student-centered education practices, highlighting the transformative impact of adaptive learning technologies in overcoming educational challenges and fostering a supportive learning environment for all students.

10. Continued improvements and future directions

As instructors, our team is dedicated to continually refining and expanding the use of Realizeit in additional Spanish courses, showcasing our commitment to ongoing improvement and innovation in educational delivery. In fact, after the launch of PAL and OER in our elementary-level Spanish courses, we collaborated with colleagues to redesign intermediate and advanced Spanish courses using the same approach. To date a total of seven Spanish courses at UCF incorporate PAL and OER, including courses that are required for the Spanish Major and Minor. In collaboration with UCF's Center for Distributed Learning and Pegasus Innovation Lab we monitor Realizeit's impact through data analysis, ensuring the tool effectively meets student needs and promotes positive educational outcomes. These are data in real time that allow us to modify course content and delivery at any given point in the semester to better serve our students. We can instantly see what is working and is not working, from a micro to a macro level, from individual student to individual section, and sometimes across multiple sections of the same course.

To build on these achievements, several recommendations and future directions are proposed:

10.1. Refine and expand course content

Continuously update the adaptive learning platform and course materials based on student feedback and performance data. These updates can occur throughout the semester in response to student inquiries, newly identified content needs, and ongoing review by the instructional team.

10.2. Broaden PAL and OER implementation

Extend the use of PAL and OER to upper-level Spanish courses, and other languages and disciplines to enhance learning across the university.

10.3. Increase faculty support and training

Provide more training and resources for faculty to effectively integrate PAL and OER, ensuring consistent quality across courses. So far, quality control has relied on support from the university's CRI and the instructional team's ongoing content review. As more instructors have joined the initiative, the team has been able to dedicate more time to improving and streamlining course materials.

10.4. Build collaborative capacity

With growing adoption of redesigned courses, the instructional team can expand, allowing for more efficient and continuous content revision.

10.5. Conduct longitudinal studies

Track the long-term impact of PAL and OER on student success to inform future improvements and ensure sustainability.

In many cases, improvement or continued development of ideas and goals does not continue. Time and funding run out. People quickly latch onto a new methodology or idea. For us, this was not a passing phase or a flashy new trend. This process was years in the making, based on our experiences and ideas about pervasive issues in language learning to the ever-changing needs of our students. Fostering collaboration among faculty, administrators, instructional designers, subject matter librarians, and others will continue to be vital for innovative course designs and culturally relevant content. Finally, sharing the insights we've gained, and our findings and key takeaways can promote the wider adoption of PAL and OER, advocating for adaptive learning technologies and open educational practices to contribute to a more inclusive and equitable education system.

11. Conclusion

The successful integration of PAL and OER in UCF's first-year Spanish course sequence has proven transformative for students in these classes and the faculty teaching them, modifying students' learning experience and enabling the team to thrive in our role as instructors. By addressing diverse student needs, enhancing engagement, and removing financial barriers, this initiative has set a benchmark for innovative teaching practices. Johar et al. (2023) indicate, "By applying learning analytics to measure student engagement in online learning, instructors and students will benefit in their teaching and learning process, respectively, to achieve academic success." We agree with this observation, and the instructional team continues to refine and expand our team's approach while remaining committed to fostering an inclusive and dynamic educational environment where every student can achieve course outcomes and their learning goals can be met.

References

Affordable Learning LOUISiana. (n.d.). *Open educational resources.* Retrieved May 5, 2025, from https://louisiana.libguides.com/oer

Center for Distributed Learning, University of Central Florida. (n.d.). Textbook affordability survey [Unpublished raw data].

Colvard, N. B., Watson, C. E., & Park, H. (2018). The impact of open educational resources on various student success metrics. *International Journal of Teaching and Learning in Higher Education, 30*(2), 262–276. https://eric.ed.gov/?id=EJ1184998

Dörnyei, Z. (2009). The L2 motivational self system. In Z. Dörnyei & E. Ushioda (Eds.), *Motivation, language identity and the L2 self* (pp. 9–42). Multilingual Matters.

Doughty, C., & Williams, J. (Eds.). (1998). *Focus on form in classroom second language acquisition.* Cambridge University Press.

Dziuban, C., Moskal, P., Parker, L., Campbell, M., Howlin, C., & Johnson, C. (2018). Adaptive learning: A stabilizing influence across disciplines and universities. *Online Learning, 22*(3), 7–39. https://doi.org/10.24059/olj.v22i3.1465

Eau, G., Judah, K., & Shahid, H. (2019). *How can adaptive platforms improve student learning outcomes? A case study of open educational resources and adaptive learning platforms.* SSRN. https://doi.org/10.2139/ssrn.3478134

Ellis, R. (2005). Measuring implicit and explicit knowledge of a second language: A psychometric study. *Studies in Second Language Acquisition, 27*(2), 141–172. https://doi.org/10.1017/S0272263105050096

El-Sabagh, H. A. (2021). Adaptive e-learning environment based on learning styles and its impact on development students' engagement. *International Journal of Educational Technology in Higher Education, 18*(1), Article 53. https://doi.org/10.1186/s41239-021-00289-4

Florida Office of Economic and Demographic Research. (2024). *Econographic news: Volume 1.* Retrieved April 14, 2025, from http://edr.state.fl.us/Content/population-demographics/reports/econographicnews_2024_Volume%201.pdf

Hilton III, J. L., Robinson, T. J., Wiley, D., & Ackerman, J. D. (2014). Cost-savings achieved in two semesters through the adoption of open educational resources. *The International Review of Research in Open and Distributed Learning, 15*(2). https://doi.org/10.19173/irrodl.v15i2.1700

Johar, N. A., Kew, S. N., Tasir, Z., & Koh, E. (2023). Learning analytics on student engagement to enhance students' learning performance: A systematic review. *Sustainability, 15*(10), 7849. https://doi.org/10.3390/su15107849

Khor, E. T., & Mutthulakshmi, K. (2024). A systematic review of the role of learning analytics in supporting personalized learning. *Education Sciences, 14*(1), 51. https://doi.org/10.3390/educsci14010051

Larsen-Freeman, D., & Anderson, M. (2011). *Techniques and principles in language teaching* (3rd ed.). Oxford University Press.

Lightbown, P. M., & Spada, N. (2013). *How languages are learned* (4th ed.). Oxford University Press.

Long, M. H. (1991). Focus on form: A design feature in language teaching methodology. In K. de Bot, R. B. Ginsberg, & C. Kramsch (Eds.), *Foreign language research in cross-cultural perspective* (pp. 39–52). John Benjamins.

Nassaji, H., & Fotos, S. (2011). *Teaching grammar in second language classrooms: Integrating form-focused instruction in communicative context.* Routledge.

Schmidt, R. (1990). *The role of consciousness in second language learning. Applied Linguistics, 11*(2), 129–158.

Siemens, G. (2013). Learning analytics: The emergence of a discipline. *American Behavioral Scientist, 57*(10), 1380–1400. https://doi.org/10.1177/0002764213498851

SPARC. (2023, February 17). *Survey shows many still struggle to afford needed learning materials.* SPARC. https://sparcopen.org/news/2023/survey-shows-many-still-struggle-to-afford-needed-learning-materials/

Tinto, V. (2017). Through the eyes of students: Rethinking student retention. *International Journal of Educational Management, 31*(2), 169–183. https://doi.org/10.1108/IJEM-12-2016-0212

UNESCO. (n.d.). *Open educational resources.* Retrieved April 14, 2025, from https://www.unesco.org/en/open-educational-resources

University of Central Florida. (n.d.). *About Hispanic serving institution.* Retrieved March 19, 2025, from https://access.ucf.edu/hispanic-serving-institution/about/

U.S. Department of Education. (n.d.). *Hispanic-serving institutions (HSIs).* Retrieved March 19, 2025, from https://sites.ed.gov/hispanic-initiative/hispanic-serving-institutions-hsis/

Tomlinson, B. (2014). *Developing materials for language teaching* (2nd ed.). Bloomsbury.

VanPatten, B. (2015). *While we're on the topic: BVP on language, acquisition, and classroom practice.* ACTFL.

Chapter 12

AI and second language education: A pedagogical perspective

Oyinkansola Sodiya
University of Texas at Arlington

Abstract

Artificial Intelligence has the potential to revolutionize second language education. However, its efficiency lies in the ability of educators to use the tools as a complementary resource rather than as a replacement for traditional teaching. This chapter argues that educators should take a systematic and theoretically grounded approach that maintains pedagogical integrity when integrating AI into second language education. Specifically, the analysis discusses that artificial intelligence tools should be integrated into phases of established instructional design frameworks such as the ELED and ADDIE models. This chapter proceeds to highlight that while AI-enhanced second language education comes with benefits, it also brings challenges such as reduced learner autonomy and algorithmic overreliance. The discussion emphasizes the value of balancing technological advances of AI with the irreplaceable role of human educators in second language education.

Keywords: Artificial intelligence, Second language education, Instructional design, Language learning pedagogy

* * *

1. Introduction

The incorporation of artificial intelligence (AI) in education has revolutionized how students engage with material and how instructors disseminate knowledge. AI, defined as technology capable of mimicking human cognitive functions (Nilsson, 1971), has been an area of interest for scholars (Maniere et al., 2023; Song et al., 2023; Ulfa, 2023). Educators now use artificial intelligence platforms to perform tasks such as generating assignments, grading, and providing feedback. Such platforms leverage technologies such ML to analyze information, identify patterns, and make predictions that support teaching effectiveness. In second language learning, artificial intelligence continues to transform how languages are taught and learned. Mastering a new language can be complex. Beyond understanding the vocabulary, a learner often also needs to develop speaking, listening, reading, and writing skills. Frequently, language learning also entails learning unique aspects, such as slang. Traditional learning methods often use repetition. While this can be helpful in learning grammar, it often limits the dynamic use of the new language. AI technologies can support this learning process. For example, using "cobots," robots collaborate with teachers to teach students key aspects of language learning, such as pronunciation (Timms, 2016). With the incorporation of AI technologies, learners can now be more engaged in the learning process and receive more frequent personalized feedback. Beyond the benefits of personalized, adaptable learning, AI technologies also offer greater flexibility in the learning process. Learners now have access to on-demand content, can practice in a judgment-free environment, and even experience simulated virtual environments.

This chapter presents a central thesis: to optimize the pedagogical value of AI in second language education, it is integral to bridge instructional design frameworks with AI technologies. It provides a comprehensive analysis of the role of AI technologies, such as intelligent tutoring systems (ITS), in revolutionizing language instruction. It also emphasizes the need to create learning environments that balance effective teaching strategies with learner autonomy. The goals for this chapter are threefold: (1) to provide evidence-based best practices for artificial intelligence integration in second language education, (2) to analyze the educational outcomes of using AI and emphasize the value of instructional design frameworks, and (3) to empower learners and educators to embrace emerging technologies such as artificial intelligence as a complementary tool in second language education rather than as a replacement. This chapter argues that AI is not simply a set of tools used in language education; it is a transformative force that requires re-addressing pedagogical frameworks in second language education.

2. Evolution of AI in language education

Before introducing technology into language education, teachers relied solely on human effort to facilitate learning. While such efforts benefitted many, they often failed to account for students' diverse needs. Early developments in artificial intelligence focused on replicating aspects of human cognition (Ulfa, 2023). There was interest in having AI technologies act, think, and solve problems like humans do. Early work on AI laid the foundation for its application across various fields. In language education, technology focuses more on grammar correction. AI tools also continued to evolve, incorporating natural language processing (NLP) and machine learning (ML) to provide more interactive and immersive learning experiences (Minire et al., 2022).

Over time, more complex systems, such as ITS, were designed to provide more personalized instruction to learners (Manire et al., 2022). Utilizing artificial intelligence technologies such as ML and NLP, ITS were designed to simulate human one-on-one tutoring experiences. The use of AI in language learning traces back to work on ITS in the 1980s (Kannan & Munday, 2018). ITS offers key benefits in second language learning. These systems create opportunities for personalized and interactive learning. They provide a significant advantage over human tutors because ITS allow learners to repeat and practice indefinitely (Kannan & Munday, 2018). These systems also perform crucial functions such as grading and providing feedback. Other popular adaptive learning platforms for second-language learning include Duolingo and Babbel.

Recent research has continued to emphasize AI's impact on language learning. Through real-time feedback and adaptive content, AI technologies improve pronunciation, enhance learner autonomy, and motivation (Rusmiyanto et al., 2023). Following an analysis of 516 papers between 2000 and 2019, Huang et al. (2023) discuss the value of the increased opportunities provided by AI. For example, educators can now use virtual reality to adopt scenario-based simulation in language practice. From a young learner's perspective, Jeon (2024) explains that chatbots reduced anxiety among Korean primary school learners by creating a judgment-free learning environment. Technology-enhanced language learning also reduces public speaking anxiety among EFL learners (Chen, 2024). Indeed, AI has advanced significantly in education.

3. Artificial intelligence foundations in second language education

3.1. Natural language processing and language interaction

One of the foundational components of artificial intelligence in second language education is natural language processing (NLP). NLP "addresses the generation of written and spoken language" (Schmidt & Strasser, 2022, p. 167). In language education, intelligent tutoring language systems (ITLS) often incorporate natural language processing technologies to generate and understand text. NLPs significantly expanded the capabilities of technology in language education. For example, learners can access an NLP platform such as ChatGPT to learn new words in a different language. NLPs are also present in virtual language tutors. Technologies like virtual language tutors and ChatGPT can simulate human conversation and provide real-time feedback.

A unique impact of natural language processing in language learning is that it has extended beyond the correctness of text and now incorporates affective elements. Through sentiment analysis, natural language processing can now detect emotions such as anger and frustration. For example, if a learner responds with a statement like "This language is hard to learn; it's so frustrating," the technology can understand the sentiment and respond empathetically with an encouraging message. This unique capability bridges a longstanding gap in second language education, as the technology can now provide learning support without undermining the value of emotional cues in learning acquisition.

While NLPs offer benefits for second-language learning, they are not without flaws. Long's (1996) interaction hypothesis emphasizes the role of the linguistic environment in second language education. While NLP tools attempt to mimic human interaction, they lack the nuances of in-person conversation, such as facial expressions and body language. Thus, technologies such as natural language processing can oversimplify the second language learning process, posing a risk to second-language education.

3.2. Adaptive learning and language learning personalization

A key application of AI in second language learning is adaptive learning. Adaptive learning gathers data from the student's learning behavior and makes a personalized learning plan that presents learning content based on the learner's profile (Cui et al., 2018). Rosetta Stone Classroom, for example, is tailored for language education settings. Adaptive learning offers an opportunity for effective blended learning, as teachers can have students alternate between classroom activities and Rosetta Stone exercises. Through Rosetta Stone Classroom's immersive learning approach, students can learn various languages with progress monitoring and customization.

Adaptive learning is well aligned with Krashen's (1982) input hypothesis. Through the concept of $i + 1$, Krashen explains that learners must be exposed to language structures they have not yet acquired. Adaptive learning pushes learners beyond their comfort level by exposing them to a variety of contexts. This technology can identify where the learner is making errors, make modifications to the learning material, and even present differing scenarios that introduce varying structures.

The promise of adaptive learning and personalization can be challenging in second-language education. Algorithmically predefined code drives personalization in adaptive learning. This pedagogical approach may limit learners' autonomy, as they encounter only scenarios accounted for in the adaptive learning technology's code.

3.3. Machine learning and language learning feedback

ML refers to systems that "derive information from or learn based on experience" (Schmidt & Strasser, 2022, p. 167). Think of a student learning Spanish on Duolingo. Using ML, Duolingo can adjust the difficulty of exercises based on a student's progress. Deep learning, on the other hand, is a subset of machine learning that involves training data to identify patterns. For example, deep learning technologies can analyze a student's pronunciation relative to that of native speakers. Such technologies can also give feedback on how to improve the student's pronunciation.

ML aligns with Schmidt's (1990) noticing hypothesis, which posits that, in language acquisition, learners must first notice variations and patterns of error. Adaptive learning technologies use machine learning to identify and modify learning content based on learner behavior. This feature is beneficial in areas such as language pronunciation, where learners can compare their pronunciation with that of native speakers and make adjustments.

Like other elements of artificial intelligence in learning, ML also comes with its challenges. ML models depend on their training data. Training data may not account for all the information needed for second-language learning, such as dialectal variation. This limitation poses a risk to learning, as it can create algorithmic dependence. Learners may have false confidence that they have mastered a language when, in fact, they have only mastered the sections covered in the machine learning training data.

Through components such as NLP, ML, and deep learning, artificial intelligence has advanced second-language learning for educators and learners. Chatbots, for example, are widely used by language learners. Learners can now access innovative, personalized learning through platforms such as Babbel and Duolingo. Such platforms also have interactive elements such as videos and quizzes. This accessibility is especially beneficial as it caters to learners who prefer self-paced learning over classroom-style teaching.

Despite the benefits of AI in second-language learning, it is essential to consider its limitations when evaluating its pedagogical promise. Overreliance on artificial technologies can present challenges such as algorithmic dependence, devaluation of interpersonal learning and cues, and an inability to replicate socio-cultural nuances. The challenge is identifying the best way to integrate AI into second-language learning as a complementary tool rather than a replacement for human teaching.

4. Impact of AI-augmented second language learning on educational outcomes

Studies have shown that AI-augmented learning positively impacts overall educational outcomes. It has been found to promote autonomy (Wei, 2023), engage students (Tiwari, 2024), improve motivation (Keerthiwansha, 2018), and enable efficient classroom management for teachers (Wang, 2021).

4.1. Learner autonomy

Studies have shown that AI-augmented learning positively impacts overall educational outcomes. It has been found to promote autonomy (Wei, 2023), engage students (Tiwari, 2024), improve motivation (Keerthiwansha, 2018), and enable efficient classroom management for teachers (Wang, 2021).

Software such as Duolingo can transcribe and correct speech. Beyond this, such software can help learners set goals and reflect on their learning. Through this, the incorporation of AI into second language learning can positively impact learners' self-regulation (Kannan & Munday, 2018). In a similar vein, adopting AI technologies in second-language education can affect learner autonomy. As AI technologies allow learners to decide when to complete lessons and how they would like to study, this can foster a sense of autonomy and independence.

4.2. Learner engagement

Engaging learning materials support second-language education. By leveraging AI-integrated technologies, educators can create learning environments that are engaging and dynamic for learners (Tiwari, 2024). For example, platforms like Duolingo and Babbel use gamification techniques to incorporate engaging features such as leaderboards and rewards. Such techniques can help attract and keep the learner's attention while learning a new language. In a study exploring AI-augmented language learning among EFL students, Underwood (2017) revealed that speaking to AI assistants was engaging. Educators can leverage this engaging aspect of AI-augmented language education to improve the learning experience. For example, educators can craft creative prompts that stimulate interaction (Eager & Brunton, 2023). The educator could prompt ChatGPT to create an in-class exercise or a dialogue relevant to the classroom population.

4.3. Cultural awareness

Cultural awareness is a crucial component of second-language learning, and AI-augmented learning has contributed to it. AI-driven applications can introduce learners to cultural contexts through features such as virtual reality and interactive storytelling. For example, through Oculus VR, learners can engage in settings that mimic real-life environments. In such environments, language learners immerse themselves in situations like ordering at a restaurant or buying food at a market. These immersive experiences can help learners better understand the culture and what it would be like to speak the language with locals. Negotiating the price of a bag in a local market while speaking a new language can be challenging. However, virtual reality can replicate such specific settings. This kind of contextual learning is vital for learners.

4.4 Motivation through open learner models

Researchers have also documented how AI-augmented second-language education affects learner motivation. In open learner models, students can access their learning outcomes. Such models can motivate learners by tracking their progress (Keerthiwansha, 2018). Intelligent computer-assisted language learning (ICALL) technologies are an example of an open learner model that motivates students (Amaral, 2007). ICALL significantly transformed student-computer interaction. These ICALL technologies leverage artificial intelligence to create more engaging, dynamic, and personalized learning environments. Unlike traditional computer-assisted language learning (CALL) systems, ICALL provides customized feedback that can motivate learners. Key elements, such as interactive features and adaptive feedback, can encourage learners to progress further in their second language learning programs.

4.5. Teacher efficiency and classroom management

Beyond the support artificial intelligence provides to learners in second language education, it also enables more efficient classroom management for teachers (Wang, 2021). Artificial intelligence technologies, such as image recognition, also support teachers by making teaching assessments more convenient (Huang & Saleh, 2021). Such technologies can create exam questions and provide feedback on assignments (Li et al., 2017). Alleviating this burden from teachers allows them to focus on more humanistic care (Huang et al., 2021). Availability to nurture healthy student-teacher relationships is crucial. However, it is also essential for educators to have visibility and input in the assessment process. For example, suppose the AI technology provides feedback on homework that has been assigned. In that case, the onus should be on the educator to review, validate, and edit as needed before the AI provides feedback to the learner. Another unique benefit of artificial intelligence technologies in education is that teachers can customize the approaches and materials. Through features such as personalized learning and adaptive learning, artificial intelligence has improved the effectiveness of instruction.

For example, AI can track a student's learning activities, identify their strengths and weaknesses, and push more material that targets their weak points. The personalized attention these AI technologies provide has created avenues to better cater to individual student needs.

4.6. Managing cognitive load

Cognitive load theory addresses the mental effort required to process information (Sweller, 2010). Learning a new language can place a high cognitive load on learners, as the process involves activities such as reading, writing, and speaking. AI strategies, such as personalized feedback, can alleviate this burden. With such a reduction in cognitive load, learners can also direct more mental effort to aspects such as vocabulary retention (Feng, 2025). Song et al.'s (2023) theoretical model – virtual reality-constructivism and cognitive load (VR-CCL) – explicitly addresses how the virtual reality environment can be used as a resource to optimize foreign language learning. VR-CCL emphasizes key factors such as design simplicity, guided interactions, and balanced visual cues. The adoption of frameworks such as the VR-CCL framework can help reduce extraneous load (load from poor instructional design) while managing intrinsic load (load from material complexity; Song et al., 2023).

The impact of AI-augmented second language learning on education is evident. It has benefits such as reducing learning time and fostering cultural awareness. Its role in driving enriched learning will undoubtedly expand, paving the way for improved education outcomes.

5. Challenges for instructors and students

AI-enhanced second language education offers many advantages. From motivating learners (Wei, 2023) to improving classroom management (Wang, 2021). However, alongside these benefits come challenges that educators, learners, and other stakeholders should consider.

5.1. Technological proficiency

The challenge of technological proficiency is one that both learners and educators must combat. While AI-driven tools can enhance the educational experience for teachers and learners, their effectiveness often depends on users' technological proficiency. This contingency poses a challenge because educators and learners often have varying technological proficiency. This disparity in technological proficiency can affect the effectiveness of AI-enhanced learning methods; for example, in a class of 30 students, five have lower technological proficiency. If the teacher incorporates AI-augmented tools such as Rosetta Stone Classroom into classroom learning, the five students with limited technological proficiency could be disadvantaged. It is also essential to consider that students may have diverse backgrounds in their use of technology. Some students may come from countries where computers are not prevalent, while others may be well-versed in numerous technology tools.

Some countries are not technologically advanced or as familiar with technology tools as the United States. Learners from such countries without experience using emerging technologies may face challenges when navigating artificial intelligence tools. This technological proficiency challenge is not exclusive to learners. Keeping up with technological advancements can also become arduous for some educators. Beyond the time investment, there are financial implications as well. Continuous learning can sometimes come at a high cost, posing a challenge for educators. The gap between educators who use and those who do not use AI in second language education can raise concerns. Are some students at a disadvantage because their teachers do not utilize artificial intelligence tools?

Addressing the challenge of technological proficiency lies in the collective effort of educators, learners, government, organizations, and AI tool manufacturers. Given the benefits AI tools bring to second-language education, some learner populations could be at a disadvantage if their educators do not explore AI's efficiencies. The reality is that many under-resourced populations lack access to AI tools. The onus lies with several parties, including the government and private organizations, to provide resources that better serve these populations. Additionally, AI tools should be designed with compatibility settings that support older devices. For example, certain apps quickly lose compatibility with older iPhone models. If some students in the class have older devices, they may lose out on the learning experience. Programmers of AI technologies should consider this and make efforts to be more inclusive in their compatibility requirements. By prioritizing inclusivity, educators can ensure that underprivileged populations without access to the latest phones and computers remain included in learning opportunities. Educators should also utilize online learning platforms available at minimal cost or free through their institutions. LinkedIn Learning, for example, has several courses on the use of AI in education.

5.2. Algorithm aversion

Algorithm aversion is another challenge in using artificial intelligence in second-language education. In discussions of willingness to use technologies, a crucial topic is algorithm aversion – a reluctance to use algorithms (Dietvorst et al., 2015). The topic of algorithm aversion is critical in language education because educators or learners who are averse to algorithms may be unwilling to work with AI tools in the learning process. Algorithms, the use of data to construct statistical models, have been proven highly influential in decision-making (e.g., Dawes, 1979). However, despite the notable impact and benefits of algorithms, algorithm aversion persists. Interestingly, even when people see algorithms outperform humans, they can still be averse to algorithms (Dietvorst et al., 2015). Algorithm aversion research has supported the notion that people may be reluctant to use algorithms due to a lack of decision control (e.g., Carringan et al., 2004). Educators may be concerned that incorporating AI tools could undermine their professional expertise in second-language education. Educators may also have concerns about their pedagogical autonomy. AI technologies are known for their ability to perform human-like tasks. In second-language education, AI technologies can create course content, determine pacing, and provide feedback. This capability can be a concern for educators, who may worry that technology can dictate a significant portion of the educational process.

5.3. Ethical considerations

Using AI tools in second language education raises multifaceted ethical questions. Using AI in grading could be problematic from an algorithmic bias perspective. If the AI tools make decisions based on pre-trained biased data, this can put specific populations at a disadvantage. From a data privacy perspective, concerns about the privacy of confidential data may make educators and institutions apprehensive about using AI tools. Oftentimes, AI tools require the collection of student information, and there may be concerns that this information could be compromised. To address these ethical issues, institutions should ensure they use AI tools that are transparent and explainable. Educators should understand the AI tools' grading process well and be readily available to discuss the AI-driven decision-making process. Additionally, while AI tools can be incorporated into the feedback component of language learning; it is crucial that educators be empowered to override AI-driven decisions.

5.4. Overreliance on AI

Incorporating AI in second language education also comes with challenges. One of the challenges highlighted in this chapter is the overreliance on AI. If educators become over-reliant on AI for instruction and decision-making, that can become problematic. Over-reliance on artificial intelligence in second language learning can undermine the value of the human educator. It can also lead to concerns about bias and fairness. Educators and developers must recognize that AI technologies can also reflect and perpetuate bias. Such bias can disadvantage specific populations. Human educators need override power, especially during the evaluation stage.

5.5. Autonomous skills

AI-enhanced language learning can negatively affect learners' autonomous skills. For example, a learner who relies on DeepL or Google Translate for translations could limit their ability to develop autonomous writing and paraphrasing skills (Schmidt & Strasser, 2022). Additionally, such translation software often does not support minority languages, thereby lacking a crucial component of authentic language learning (Schmidt & Strasser, 2022). In the AI era, there is a shift towards a student-centered approach, with students playing a more active role in their learning (Fu, 2020). Teachers should be cognizant as they incorporate AI-enabled learning into second-language education. In the student-centered approach, the students can adapt their lesson plans and decide what to learn. While the student approach may be optimal for some, students lacking self-regulation may be disadvantaged. Teachers should also focus on helping students cultivate skills crucial to autonomous learning, such as goal-setting and time management.

Collectively, as drivers of the educational process, educators and institutions need to be aware of the challenges of using AI tools in second language learning. By addressing these challenges head-on, academic institutions can ensure they increasingly leverage the benefits of AI effectively in second-language education.

6. Best practices for AI-enhanced language education

6.1. Adopt an instructional design framework

In addressing the use of AI in second-language education, an effective way to optimize the pedagogical value of AI is to critically reflect on instructional design models and identify phases that are optimal for AI integration. Two instructional design frameworks worth considering here are the eLlearning engagement design (ELED) framework (Czerkawski & Lymann, 2016) and the analysis, design, development, implementation, and evaluation (ADDIE) model (Seel & Glasgow, 1998). These frameworks outline structured phases for integrating AI tools. Adopting an instructional design framework that fosters learner engagement could enable educators to fully harness the potential of artificial intelligence in second language education. To ensure that the integration of AI in second language learning is pedagogically grounded, this chapter posits that AI integration should align with specific instructional design frameworks.

eLearning engagement design framework. The eLearning engagement design (ELED) framework consists of four phases: identifying instructional needs, defining instructional goals, developing learning environments, and evaluating instructional effectiveness. The first phase of determining the learner's instructional needs is critical in AI-driven second language learning. In this phase, educators can use artificial intelligence tools to identify learners' needs and pinpoint areas with skill gaps. For example, while one student may need to learn speaking skills in the second language, another may be conversational and need to improve their writing skills. AI-driven assessments can be instrumental in identifying a learner's proficiency. In this phase, educators can incorporate AI tools such as Speechace and ReadTheory to analyze learners' fluency and proficiency. Educators can better tailor their teaching materials to meet individual needs by drawing on insights into students' instructional needs. This personalized approach is crucial for student engagement and can help identify students needing additional support.

The subsequent phases of the ELED framework also offer unique opportunities to incorporate AI into second language education. Phase 3 – developing a learning environment, for example, provides educators with opportunities to integrate artificial intelligence technologies. Mondly VR and ImmerseMe are examples of virtual reality language-learning platforms that offer unique approaches to second-language learning. On these platforms, learners can immerse themselves in virtual environments to practice the target language. Not only does this support grammar learning, but it also enhances the learner's conversational skills in the target language.

The ADDIE model. Another popularly accepted model—the ADDIE model—consists of five stages: analysis, design, development, implementation, and evaluation. In the analysis stage, the learning objectives and needs are identified. In the analysis phase, educators can use artificial intelligence tools to identify learner profiles. For example, by analyzing the learner's text history, the AI tool can highlight repeated errors that the learner may not be aware of. Through this, the tool can signal the need for scaffolding in certain areas. In the design phase, educators outline the overall learning experience, including content layout and assessment tools. In the development phase, the course content is developed and revised. Here, teachers can use AI platforms like Canva to create engaging content. This type of content would not only be visually appealing but would also help teachers combine verbal and visual learning materials – providing support for Johnson and Mayer's (2009) multimedia learning effect. During the implementation phase, educators deliver instructional materials to learners. Here, teachers can use platforms like Duolingo to supplement coursework. Such variation during the implementation phase can also help educate learners about the nuances of second-language learning, such as pitch. In the evaluation phase, a formal assessment of the entire process is conducted (Seel & Glasgow, 1998). Here, educators can use AI dashboards to access student grades and identify learning patterns.

The integration of AI into instructional design frameworks varies across learning levels. For example, a K-12 teacher may prioritize adopting AI during the implementation phase to create engaging content that captures students' attention. On the other hand, a university professor may prioritize implementing AI in the design phase of second-language learning to allocate more class time to in-person feedback and discussions with the professor.

Incorporating AI into second-language education does not necessarily entail replacing traditional learning with AI technologies. Educators can integrate AI technologies into specific phases of instructional design frameworks, such as the ELED and ADDIE models. Pedagogical goals and instructional design frameworks are crucial. By leveraging instructional design frameworks such as the ADDIE and ELED (models, educators can ensure they incorporate AI tools into the phases of learning that would be most beneficial to learners. This process could also help educators ensure they take a strategic, goal-oriented approach to the learner's education. By laying out the phases of the teaching process, educators can decipher which phases of second language education would benefit from leveraging artificial intelligence competencies. There is no universal answer here, as different audiences have different pedagogical goals. A K-12 language educator, for example, may emphasize AI during the implementation phase of the ADDIE model to prioritize engaging, fun learning content for children. On the other hand, a college language educator may prioritize incorporating AI into the design phase of language learning to allocate more time to giving students feedback. This process can serve as a reflective moment for educators to take a step back and analyze how best to leverage AI in their teaching.

6.2. Understand the value of human educators

Language educators should leverage artificial intelligence and move beyond routine administrative tasks. AI-driven insights have opened a world of opportunity, and educators should not fall behind in utilizing AI efficiencies. ITS, for example, have been recorded to have many benefits for learners. It is also important not to undermine the value of human educators. In a meta-analysis of 39 studies, Steenbergen-Hu and Cooper (2014) found that ITS outperformed all other instruction methods, they were still less effective than human tutoring. These findings highlight the importance of human educators. Thus, this chapter emphasizes that while artificial intelligence offers many benefits in second language education, educators must preserve the human touch.

While AI features like sentiment analysis have helped AI technologies detect and respond to emotion, they cannot respond to emotional cues like a human would. Affect also comes into play as some students have their teachers as their role models. Getting encouragement from a teacher one admires is not the same as receiving encouragement from an AI tool.

6.3. Balance integration and human instruction

Several studies (e.g., Carringan et al., 2004; Scherer et al., 2015) have identified a lack of decision control as one of the leading causes of algorithm aversion. If educators feel they are not involved in decision-making, they may become more resistant to using AI technologies in their teaching. Thus, it is crucial to ensure that AI does not undermine the role of human educators. Keeping humans in the loop helps address algorithm aversion (Burton et al., 2020). Having humans in the loop has been identified as a solution to algorithm aversion (Burton et al., 2020). AI tools should be introduced to complement, rather than replace, human educators in second-language learning. Though AI makes commendable attempts to mimic human minds, it still cannot replicate meaningful human connections. For example, human educators can bring emotion, empathy, and intuition into the classroom. These distinct qualities also help nurture meaningful human relationships.

AI technologies such as natural language processing and machine learning have created dynamic, adaptive learning environments in ways previously unimaginable. Research and practice have documented a shift toward more personalized second-language learning. This shift emphasizes that learning a second language is not a one-size-fits-all approach. Educators can now tailor curricula and teaching methods to meet learners' specific needs.

7. Looking ahead

As AI continues to advance, its role in language learning grows increasingly sophisticated, dynamic, and interconnected with broader pedagogical, cultural, and ethical dimensions. Three interrelated developments characterize the next phase of AI-enhanced language learning: improved functionalities, stronger cultural considerations, and a deeper focus on responsible AI use.

7.1. Improved functionalities

As AI continues to evolve, its impact and applications in language learning are likely to expand. The future of AI-enhanced language learning is promising, as scholars and practitioners are working to address its current challenges and enhance its capabilities. For example, in the Duolingo app, users can now access a premium feature that lets them have a video call with an AI character—Lily (Seitz, 2024).

7.2. Improved cultural considerations

Research has shown that cultural considerations are essential in second-language learning. Scollon and Scollon (1995), for example, discuss how imposing one's opinion could be regarded as impolite in some Asian cultures. Using words like "clearly" or "undoubtedly" may not be well received in such cultures. Likely, AI tools will continually account for such cultural considerations in second-language learning, primarily as conversational agents are intended to mimic human communication.

7.3. Responsible AI

As educators continue to incorporate AI tools into education, discussions about ethics will likely persist. Organizations and scholars should continue investigating factors such as data privacy and algorithmic bias. Institutions should also have adequate training programs in place to ensure educators are cognizant of these ethical considerations. There is also likely to be more government involvement as companies tackle concerns about data retention. For example, critics have expressed concerns about education systems that store the facial expressions of minors in China (Nesta, n.d.).

8. Conclusion

The integration of AI into second language education has transformative potential, but this lies in adopting thoughtful, theoretically grounded approaches. This chapter has emphasized that while AI-enhanced second-language learning has benefits, it also comes with challenges, such as algorithmic overreliance, poor training data, and reduced learner autonomy. AI tools must enhance, rather than dictate, the learning process for students. Scholars, educators, and learners must recognize and preserve the importance of relational teaching and the nuances that emerge through human-to-human interactions.

The goal is for second language educators to work collaboratively with AI tools in ways that enhance, rather than dilute, the pedagogical integrity of second language learning. By integrating AI into learning frameworks such as the ELED and ADDIE models, educators can approach AI use systematically, practically, and theoretically grounded. By taking a theoretically grounded approach to embracing artificial intelligence, learners and educators can navigate the complexities of language learning in this rapidly evolving technological age. With thoughtful implementation, AI has the potential to revolutionize language education, empowering both learners and teachers to achieve an engaging, efficient experience.

References

Amaral, L. A. (2007). *Designing intelligent language tutoring systems for integration into foreign language instruction* (Publication No. AA13262074) [Doctoral dissertation]. The Ohio State University. ProQuest Dissertations and Theses Global.

Burton, J. W., Stein, M. K., & Jensen, T. B. (2020). A systematic review of algorithm aversion in augmented decision making. *Journal of Behavioral Decision Making, 33*(2), 220-239. https://doi.org/10.1002/bdm.2155

Carrigan, N., Gardner, P. H., Conner, M., & Maule, J. (2004). The impact of structuring information in a patient decision aid. *Psychology & Health, 19*(4), 457-477. https://doi.org/10.1080/08870440310001652641

Chen, Y. C. (2024). Effects of technology-enhanced language learning on reducing EFL learners' public speaking anxiety. *Computer Assisted Language Learning, 37*(4), 789-813. https://doi.org/10.1080/09588221.2022.2055083

Cui, W., Xue, Z., & Thai, K. P. (2018, November). Performance comparison of an AI-based adaptive learning system in China. In 2018 Chinese Automation Congress (CAC; pp. 3170-3175). IEEE. https://doi.org/10.1109/CAC.2018.8623327

Dawes, R. M. (1979). The robust beauty of improper linear models in decision-making. *American Psychologist, 34*(7), 571–582. https://doi.org/10.1037/0003-066X.34.7.571

Dietvorst, B. J., Simmons, J. P., & Massey, C. (2015). Algorithm aversion: People erroneously avoid algorithms after seeing them err. *Journal of Experimental Psychology: General, 144*(1), 114-126. https://doi.org/10.1037/xge0000033

Eager, B., & Brunton, R. (2023). Prompting higher education towards AI-augmented teaching and learning practice. *Journal of University Teaching and Learning Practice, 20*(5). https://doi.org/10.53761/1.20.5.02

Feng, L. (2025). Investigating the effects of artificial intelligence-assisted language learning strategies on cognitive load and learning outcomes: A comparative study. *Journal of Educational Computing Research, 62*(8), 1961-1994. https://doi.org/10.1177/07356331241268349

Fu, Y. (2020). Research on the development trend of the online education industry considering the influence of big data and artificial intelligence. In Z. Xu, K. K. Choo, A. Dehghantanha, R. Parizi, & M. Hammoudeh (Eds.), *Cyber security intelligence and analytics* (pp. 852-859). Springer International. https://doi.org/10.1007/978-3-030-15235-2_114

Huang, J., Saleh, S., & Liu, Y. (2021). *A review on artificial intelligence in education. Academic Journal of Interdisciplinary Studies, 10*(3). 206-217. https://doi.org/10.36941/ajis-2021-007

Huang, X., Zou, D., Cheng, G., Chen, X., & Xie, H. (2023). Trends, research issues and applications of artificial intelligence in language education. *Educational*

Technology & Society, 26(1), 112-131. https://doi.org/10.30191/ETS.202301_26(1).0009

Jeon, J. (2024). Exploring AI chatbot affordances in the EFL classroom: Young learners' experiences and perspectives. *Computer Assisted Language Learning, 37*(1-2), 1-26. https://doi.org/10.1080/09588221.2021.2021241

Johnson, C. I., & Mayer, R. E. (2009). A testing effect with multimedia learning. *Journal of Educational Psychology, 101*(3), 621-628. https://doi.org/10.1037/a0015183

Kannan, J., & Munday, P. (2018). New trends in second language learning and teaching through the lens of ICT, networked learning, and artificial intelligence. *Círculo De Lingüística Aplicada a La Comunicación, 76*(76), 13-30. https://doi.org/10.5209/CLAC.62495.

Keerthiwansha, N. W. B. S. (2018). Artificial intelligence education (AIEd) in English as a second language (ESL) classroom in Sri Lanka. *Artificial Intelligence, 6*(1), 31-36.

Krashen, S. (1982). *Principles and practice in second language acquisition.* https://doi.org/10.2307/3586656

Li, Q., Cao, H., & Lu, Y. (2017, November). Connecting paper to digitization: A homework data processing system with data labeling and visualization. In *Proceedings of the 14th EAI International Conference on Mobile and Ubiquitous Systems: Computing, Networking and Services* (pp. 504-510). Networking and Services. Association for Computing Machinery. https://doi.org/10.1145/3144457.3152353

Long, M. H. (1996). The role of the linguistic environment in second language acquisition. In W. C. Ritchie, & T. K. Bhatia (Eds.), *Handbook of second language acquisition* (pp. 413-468). Academic Press.

Nesta. (n.d.). AI applications in education: Privacy and accessibility. In China's approach to AI ethics. Retrieved January 14, 2025, from /

Nilsson, N. J. (1971). *Problem-solving methods in artificial intelligence.* McGraw-Hill.

Rusmiyanto, R., Huriati, N., Fitriani, N., Tyas, N. K., Rofi'i, A., & Sari, M. N. (2023). The role of artificial intelligence (AI) in developing English language learner's communication skills. *Journal on Education, 6*(1), 750-757. https://doi.org/10.31004/joe.v6i1.2990

Scherer, L. D., de Vries, M., Zikmund-Fisher, B. J., Witteman, H. O., & Fagerlin, A. (2015). Trust in deliberation: The consequences of deliberative decision strategies for medical decisions. *Health Psychology, 34*(11), 1090-1099. https://doi.org/10.1037/hea0000203

Schmidt, R. W. (1990). The role of consciousness in second language learning. *Applied Linguistics, 11*(2), 129-158. . http://dx.doi.org/10.1093/applin/11.2.129

Schmidt, T., & Strasser, T. (2022). Artificial intelligence in foreign language learning and teaching: A CALL for intelligent practice. *Anglistik: International Journal of English Studies, 33*(1), 165-184. https://doi.org/10.33675/ANGL/2022/1/14

Scollon, R., & Scollon, S. B. K. (1995). *Intercultural communication: A discourse approach.* Blackwell.

Seel, B., & Glasgow, Z. (1998). *Making instructional design decisions.* Merrill.

Seitz, P. (2024, September 24). *Duolingo adds AI innovations to language learning app.* Investor's Business Daily. Retrieved January 13, 2025, from https://www.investors.com/news/technology/duolingo-stock-ai-innovations-language-learning-app/

Song, C., Shin, S. Y., & Shin, K. S. (2023). Optimizing foreign language learning in virtual reality: A comprehensive theoretical framework based on constructivism and cognitive load theory (VR-CCL). *Applied Sciences, 13*(23), 12557. https://doi.org/10.3390/app132312557

Steenbergen-Hu, S., & Cooper, H. (2014). A meta-analysis of the effectiveness of intelligent tutoring systems on college students' academic learning. *Journal of Educational Psychology, 106*(2), 331. https://doi.org/10.1037/a0034752

Sweller, J. (2010). Element interactivity and intrinsic, extraneous, and germane cognitive load. *Educational Psychology Review, 22,* 123-138. https://doi.org/10.1007/s10648-010-9128-5

Timms, M. J. (2016). Letting artificial intelligence in education out of the box: Educational cobots and smart classrooms. *International Journal of Artificial Intelligence in Education, 26,* 701-712. https://doi.org/10.1007/s40593-016-0095-y

Tiwari, H. P. (2024). Artificial intelligence in the classroom: Revolutionizing English language teaching. *Journal of English Teaching and Linguistics Studies, 6*(1), 42-59. https://doi.org/10.55215/jetli.v6i1.9757

Underwood, J. (2017). Exploring AI language assistants with primary EFL students. In K. Borthwick, L. Bradley, & S. Thouësny (Eds.), *CALL in a climate of change: Adapting to turbulent global conditions – Short papers from EUROCALL 2017* (pp. 317-321). Research-publishing.net. https://doi.org/10.14705/RPNET.2017.EUROCALL2017.733

Wang, Y. (2021). An improved machine learning and artificial intelligence algorithm for classroom management of English distance education. *Journal of Intelligent & Fuzzy Systems, 40*(2), 3477-3488. https://doi.org/10.3233/JIFS-18938

Wei, L. (2023). Artificial intelligence in language instruction: Impact on English learning achievement, L2 motivation, and self-regulated learning. *Frontiers in Psychology, 14,* 1261955. https://doi.org/10.3389/fpsyg.2023.1261955

Contributors

Nebahat Badem is an Assistant Professor at Kahramanmaras Sutcu İmam University in Kahramanmaras, Turkey. She holds a Ph.D. in ELT from Cukurova University in Adana, Turkey. Her main research interest is language in cognition. Recently, she has been intrigued by the use of AI in language education.

Rebecca Bias is Assistant Director of the Center for Languages, Literatures and Cultures and Director of the Collegiate Assessment of Academic Proficiency (CAAP) Program at Ohio State University (OSU). CAAP offers early language assessment for high school students in Spanish, French, German, and Chinese. She earned her Ph.D. in French Cinema and Technology in Language Learning from OSU in 2005. Dr. Bias coordinates career events, oversees graduate training, manages scholarships, and mentors sophomores through STEP. She is currently researching virtual reality headset use in introductory language and culture courses, aiming to enhance student engagement and learning outcomes through immersive technology.

Samanta Buffa is a PhD student and graduate teaching assistant in Italian in the Department of French and Italian at The Ohio State University. Her dissertation is a comparative study of the use of virtual reality in the Italian classroom informed by her expertise in instructed second language acquisition and world language pedagogy. She has presented her research at a variety of conferences, and, in 2024, she received the Graduate Associate Teaching Award, Ohio State's highest recognition of the exceptional teaching provided by a graduate teaching assistant.

Fang Cao is affiliated with Seoul National University, where she earned a master's degree in Educational Technology and completed doctoral coursework in Education. She serves as an AI course consultant for Chinese Plus and is a member of the International Chinese Teachers' Digital Literacy Expert Pool of the Ministry of Education's Center for Language Education and Cooperation (CLEC), China. Her work focuses on teacher digital literacy, AI-generated content, and professional development for international Chinese language teachers.

Wing Shan Chan is an independent scholar affiliated with the National Coalition of Independent Scholars (NCIS). She holds a PhD in Linguistics from Macquarie University, where she explored the cognitive impact of subtitles in higher education using eye-tracking technology. Fluent in Cantonese, English, and Mandarin, she brings a multicultural perspective to her work. Her professional journey spans research consultancy, higher education, library services, and therapeutic arts, fields that demonstrate her commitment to empathetic engagement and the responsible use of knowledge. With interdisciplinary expertise in scholarly communication, literature synthesis, and information management, she empowers others through reflective inquiry and compassionate service.

Baiyun Chen is the Program Director for the Personalized Adaptive Learning team at the Center for Distributed Learning at the University of Central Florida (UCF). She earned her PhD in Instructional Technology and Design from the UCF, and she has published widely on topics related to online and adaptive instruction, including 23 peer-reviewed journal articles, book chapters, and editorial-reviewed articles. She has also delivered more than 50 presentations at international and local conferences since 2006.

Su-Je Cho is a Professor of Special Education and Chair of the Division of Curriculum and Teaching at Fordham University's Graduate School of Education. She has worked in the field of special education for more than 30 years as a teacher, professor, researcher, and advocate for children with disabilities and their families. Her research focuses on training educators and families in assessment and interventions for students with challenging behaviors. Dr. Cho has directed multiple grant projects funded by the New York State Department of Education and the U.S. Department of Education's Office of Special Education Programs. She is also a recipient of Fordham University's Outstanding Externally Funded Research Award on Interdisciplinary Research.

Xiaoqing Gu is a Professor and Ph.D. supervisor at East China Normal University, where she serves as the Chair of the Department of Educational Information Technology and the Director of the Shanghai Engineering Research Center for Digitalized Education Equipment. She has led numerous national and international projects, publishing over 200 papers in high-impact journals and authoring more than 20 books and textbooks. She is the Editor-in-Chief of *International Journal of Smart Technology and Learning* and Co-Editor-in-Chief of *npj Science of Learning*. Her recent research focuses on digital education, the role of AI in empowering education, and learning analytics technologies.

Ömer Buğra Hamzaoğlu is a highly accomplished, creative and artistic graduate student in the field of English Language Teaching, with a degree in Physics from Middle East Technical University, one of the top universities in Turkey. His interdisciplinary background in languages, arts, and physics allows him to combine his unique scientific thinking with innovative language research techniques. His research fields of interest include Applied Linguistics, AI Integration in Education, Language Acquisition, and the development of Adaptive Gaming and Learning Technologies. He is especially interested in modernizing language teaching methods for his upcoming research studies in the field of languages.

Başak Eda Hancı-Azizoğlu gained her PhD in English literature from Indiana State University of Pennsylvania with a double major in rhetoric and applied linguistics/TESOL. She is an expert on global perspectives on language and literacy policies and English language learning and writing programs. She also has delivered master's level courses in leadership and language teaching methods to certified American school teachers who teach multicultural students. She is currently working as an Associate Professor in the English Language Teaching Department at Mediterranean University, Türkiye. Her research interests are artificial linguistic intelligence, linguistic analysis, and the poetic and creative function of English as an additional language.

Belle Li is a PhD candidate in learning design and technology at Purdue University. Her research investigates how AI and digital technologies transform self-directed learning, epistemic cognition, and assessment. Specifically, she explores generative AI and other emerging technologies as tools to support learners' motivation, autonomy, and strategy use. Her publications span topics such as AI-mediated learning environments, cognitive offloading, and learner autonomy in human-AI interaction. Li's long-term goal is to advance research at the intersection of learning sciences, AI, and human-computer interaction, contributing to theory while designing practical systems that promote equity, autonomy, and meaningful learning.

Xiaoxuan Li is a postdoctoral researcher at Penn State University, where she works on projects addressing teacher well-being in special education. She earned her Ph.D. in Educational Leadership and Policy Studies and an M.S. in

Foreign and Second Language Education from Florida State University. Her research explores language educator identity, learner experiences, and the integration of innovative practices to support teachers and students. Li's scholarship reflects her commitment to advancing equity, well-being, and cross-cultural perspectives in education.

Xiaying Li is a PhD candidate in curriculum and instruction and an adjunct instructor at Fordham University, as well as a public high school Chinese language teacher in New York. Her research lies at the intersection of special education, educational technology, and language acquisition, with a particular focus on preparing world language teachers to integrate emerging tools such as VR, AR, and AI to create inclusive and accessible classrooms. Drawing on her dual role as practitioner and researcher, she is committed to applying theory to practice and using classroom experience to inform theory, advancing equity, accessibility, and engagement in language education. Through her scholarship and teaching, she seeks to foster innovative, technology-enhanced approaches to inclusive language learning.

Omer Ozer is an Associate Professor at Adana Alparslan Turkes Science and Technology University, Turkey. He has published extensively in the fields of multilingual policies in higher education, computer-assisted language learning, and autonomous language learning.

Anne Prucha is a senior instructor in Modern Languages and Literatures at the University of Central Florida, where she teaches Spanish and TESOL. She focuses on enhancing student success by integrating Adaptive Learning and Open Educational Resources into Spanish instruction. Collaborating with UCF's Pegasus Innovation Lab, she works to improve accessibility and reduce costs. She is currently working on a needs analysis study to explore interest in a 360° digital asset collection in the Florida State University System. This initiative, in partnership with UCF's Center for Distributed Learning, supports curriculum internationalization and aims to develop a sample library of virtual environments for use in online learning across disciplines and global contexts.

Maria Redmon is a Senior Instructor of Spanish Language and Culture at the University of Central Florida (UCF). Her research interests and publications include enhancing learning through the innovative use of technologies such as 360° videos, universal design, augmented reality, artificial intelligence, digital storytelling, and collaborative online international learning (COIL), among others. Professor Redmon has received teaching and learning awards at UCF; she has also been awarded NSF, Fulbright-Hays, Title VI, and NEH grants for her research and curriculum development. Professor Redmon has published in international and national journals focused on teaching, innovation, and learning. She holds degrees from the Universidad Complutense de Madrid and the Madrid Chamber of Commerce and Industry, Spain. Professor Redmon initiated the Translation and Interpretation Certificate program, the Elementary Spanish Honors, Business Spanish, and High Impact courses in Latin American Studies and Business at UCF.

Kevin Richards is a Professor of Teaching and Outreach Coordinator in the Department of Germanic Languages and Literatures at The Ohio State University. He designs and teaches large-enrollment online General Education courses that examine how German culture, history, and identity intersect with innovations in technology and participatory citizenship. In addition, he collaborates with interdisciplinary teams on projects that apply emerging technologies to cultural interpretation and language learning.

SM Sabra is a student at the University of California, Irvine. A student of sociology, Sabra is committed to exploring the social, cultural, and artistic dimensions of happiness and inequality; the intersections of infrastructure, well-being, and art; and the role of social structures and creative expression in shaping happiness. With an interest in global inequality and human flourishing, Sabra is multilingual; she is fluent in English and Arabic, with varying levels of mastery in Spanish, Mandarin Chinese, and Tagalog/Filipino. Sabra aims to promote the theory that happiness is a choice. She is also a student of the arts.

Jess Smith is Assistant Professor of Literacy Education at Bellarmine University in Louisville, Kentucky. She earned her Ph.D. in Curriculum and Teaching from Baylor University. Her research explores how mentoring relationships and instructional practices shape the ways educators support reading and writing development across contexts with particular attention to equity and access. She has published and presented nationally on topics such as English language development, banned books, and writing self-efficacy. Her current projects include a multiple case study of teacher preparation for censorship, a mixed-methods study of AI in teacher education, and an exploration of teacher perceptions of fake reading.

Oyinkansola Sodiya is an adjunct Assistant Professor at the University of Texas at Arlington. Her research explores artificial intelligence augmentation in the workplace. With her industry background in robotics process automation, she has held leadership roles in companies such as Boeing Aircrafts and Raymond James Financial. Dr. Sodiya is a fellow of the Chartered Management Institute, a board member of the Southwest Academy of Management, and a track chair at the Midwest Academy of Management. In recognition of her contributions, she was honored with the 2024 United States Department of Transportation Dwight David Eisenhower Achievement Award.

Vera Sotirovska is a Clinical Assistant Professor at Seattle University. As a Fulbright scholar and a former elementary and middle school teacher, she is a passionate about conducting research in literacy education across transnational contexts. Her research publications have appeared in leading publication outlets, such as *The Elementary School Journal, Early Childhood Education Journal, Language Teaching Research, The Reading Teacher*, and *Teacher Education.* She also is the lead author of *Enhancing Educators' Theoretical and Practical Understandings of Critical Literacy*, published by Cambridge University Press; this volume offers practical guidance on incorporating critical literacy in teacher education.

Kacie Tartt is a Senior Instructor in the Department of Modern Languages and Literatures at the University of Central Florida (UCF), where she emphasizes adaptive learning and open education resources to enhance access, engagement, and student success. Her courses align with UCF's Affordable Instructional Materials initiative, for which she received the AIM High Group Award in 2022. She collaborates with the Center for Distributed Learning to improve outcomes for historically marginalized students. Her work is archived in UCF's STARS Repository. In addition to teaching and curriculum development, she co-directed a Florida Humanities grant focused on Latinx creative writing in Florida.

Jessica Tojo-Raible is an Associate Instructional Designer at the University of Central Florida's Center for Distributed Learning. She specializes in Personalized Adaptive Learning (PAL) and has presented both nationally and internationally on its implementation in online education. Her research interests focus on PAL, open educational resources, and the quality of online courses.

Shelly Wyatt is a Senior Instructional Designer at the University of Central Florida's Center for Distributed Learning. In 2013, she received her PhD in Instructional Technology and Design (IDT) at UCF, with a research focus on technology and second language acquisition. Dr. Wyatt has been an active contributor to the scholarship of IDT, presenting at conferences in the US and abroad. She has also received several awards associated with her work in IDT, including a Research Incentive Award (UCF, 2023), Quality Impact Award (UCF, Center for Distributed Learning, 2022), and Excellence in Distance Learning Research – Bronze level (United States Distance Learning Association, 2024).

Jiawen Zhu is currently a Lecturer in the Faculty of Education at East China Normal University. She received her Ph.D. in Curriculum and Instruction with an emphasis on educational technology from the University of Florida and holds a master's degree in foreign language education from the University of Pittsburgh. Her research focuses on online learning communities, multimedia learning, and the role of AI in empowering education. She serves as an Assistant Editor for the *International Journal of Smart Technology and Learning*, and a reviewer for several SSCI journals and international conferences in the field of educational technology.

Index

A

B

C

D

F

G

H

I

M

N

O

P

S

T

U

V

W

Z

A masterpiece in scholarly communication -- Relevant…clear…engaging!

Prof. Dr. Michael Simonson
Fischler College of Education and School of Criminal Justice
Nova Southeastern University

This is a great book written by second language and foreign language educators, for all educators, regardless of their discipline. […] The book bridges the gap and provides just-in-time professional development opportunities for language educators, while emerging technologies (such as Artificial Intelligence and Virtual Reality) and pedagogies (such as Personalized Adaptive Learning) keep pushing the boundaries for educators and learners and reshaping the new norms of the educational landscape. This is also a great book for students in pre-service language teacher preparation and training programs at both undergraduate and graduate levels.

Dr. Yunhong "Tom" Tu
Statewide Programs Coordinator
Florida Virtual Campus

www.ingramcontent.com/pod-product-compliance
Lightning Source LLC
LaVergne TN
LVHW010533300926
849162LV00016B/1872/J
* 9 7 9 8 2 6 1 6 0 0 5 8 9 *